Body Tracking in Healthcare

Synthesis Lectures on Assistive, Rehabilitative, and Health-Preserving Technologies

Editor

Ron Baecker, *University of Toronto*

Advances in medicine allow us to live longer, despite the assaults on our bodies from war, environmental damage, and natural disasters. The result is that many of us survive for years or decades with increasing difficulties in tasks such as seeing, hearing, moving, planning, remembering, and communicating.

This series provides current state-of-the-art overviews of key topics in the burgeoning field of assistive technologies. We take a broad view of this field, giving attention not only to prosthetics that compensate for impaired capabilities, but to methods for rehabilitating or restoring function, as well as protective interventions that enable individuals to be healthy for longer periods of time throughout the lifespan. Our emphasis is in the role of information and communications technologies in prosthetics, rehabilitation, and disease prevention.

Human Factors in Healthcare: A Field Guide to Continuous Improvement
Avi Parush, Debi Parush, and Roy Ilan
2017

Assistive Technology Design for Intelligence Augmentation
Stefan Carmien
2016

Body Tracking in Healthcard
Kenton O'Hara, Cecily Morrison, Abigail Sellen, Nadia Bianchi-Berthouze, and Cathy Craig
2016

Clear Speech: Technologies that Enable the Expression and Reception of Language No Access
Frank Rudzicz
2016

Designed Technologies for Healthy Aging
Claudia B. Rebola
2015

Body Tracking in Healthcare
Kenton O'Hara, Cecily Morrison, Abigail Sellen, Nadia Bianchi-Berthouze, Cathy Craig

ISBN: 978-3-031-00472-8   print
ISBN: 978-3-031-01600-4   ebook

DOI 10.1007/978-3-031-01600-4

A Publication in the Springer series
SYNTHESIS LECTURES ON ASSISTIVE, REHABILITATIVE, AND HEALTH-PRESERVING TECHNOLOGIES, #9
Series Editors: Ronald M. Baecker, University of Toronto

Series ISSN 2162-7258 Print 2162-7266 Electronic

Body Tracking in Healthcare

Kenton O'Hara
Microsoft Research, Cambridge
Cecily Morrison
Microsoft Research, Cambridge
Abigail Sellen
Microsoft Research, Cambridge
Nadia Bianchi-Berthouze
University College London
Cathy Craig
Queens University, Belfast

SYNTHESIS LECTURES ON ASSISTIVE, REHABILITATIVE, AND HEALTH-PRESERVING TECHNOLOGIES #9

ABSTRACT

Within the context of healthcare, there has been a long-standing interest in understanding the posture and movement of the human body. Gait analysis work over the years has looked to articulate the patterns and parameters of this movement both for a normal healthy body and in a range of movement-based disorders. In recent years, these efforts to understand the moving body have been transformed by significant advances in sensing technologies and computational analysis techniques all offering new ways for the moving body to be tracked, measured, and interpreted. While much of this work has been largely research focused, as the field matures, we are seeing more shifts into clinical practice. As a consequence, there is an increasing need to understand these sensing technologies over and above the specific capabilities to track, measure, and infer patterns of movement in themselves. Rather, there is an imperative to understand how the material form of these technologies enables them also to be situated in everyday healthcare contexts and practices. There are significant mutually interdependent ties between the fundamental characteristics and assumptions of these technologies and the configurations of everyday collaborative practices that are possible them. Our attention then must look to social, clinical, and technical relations pertaining to these various body technologies that may play out in particular ways across a range of different healthcare contexts and stakeholders. Our aim in this book is to explore these issues with key examples illustrating how social contexts of use relate to the properties and assumptions bound up in particular choices of body-tracking technology. We do this through a focus on three core application areas in healthcare—assessment, rehabilitation, and surgical interaction—and recent efforts to apply body-tracking technologies to them.

KEYWORDS

body tracking, healthcare, rehabilitation, assessment, motion tracking, technology, computer vision, computing, human-computer interaction, touchless interaction, data, algorithms, health, sensors, gait analysis, mobile, gesture, medical imaging, accelerometer, depth sensor, force sensors, inertial measurement unit, balance board, body-worn sensors, interactive technology, physiotherapy, doctor, patient, multiple sclerosis, chronic pain, stroke, fall risk, Parkinson's disease, cameras, activity monitoring, collaboration, teamwork, kinematics, kinetics, robotics, wearable computing, exergames, surgery, natural user interfaces, exercise, older adults, natural user interface, speech, movement disorder

Dedication

To Dylan and Darby

Contents

List of Figures

Figures 2.1, 2.3, and 2.5 are from an article by the authors published in *Human-Computer Interaction*, 2016 (copyright Taylor and Francis), available online at www.tandfonline.com/10.1080/07370024.2015.1093421.

Figures 2.2 and 2.6 are from an article by the authors published in *Medical Image Computing and Computer-Assisted Intervention – MICCAI 2014*, Volume 8674 of the series Lecture Notes in Computer Science, 2014 (copyright Springer), available online at http://link.springer.com/10.1007/978-3-319-10470-6_54.

Figure 3.1 is from an article by the authors published in *CHI 2014* (copyright ACM), available online at http://dx.doi.org/10.1145/2556288.2557268 and from an article by the authors published in *Human-Computer Interaction*, 2015 (copyright Taylor and Francis), available online at http://www.tandfonline.com/10.1080/07370024.2015.1085310.

Figure 3.2 is from an article by the authors published in *Games for Health Journal*, 2015 (copyright Mary Ann Liebert, Inc), available online at http://online.liebertpub.com/doi/full/10.1089/g4h.2015.0006.

Figures 4.1 and 4.6 are from an article by the authors published in *Transactions on Computer Human Interaction*, 2013 (copyright ACM), available online at http://dl.acm.org/citation.cfm?doid=2442106.2442111.

Figure 4.2 is from article by the authors published in *CHI '11, 2011*, (copyright ACM), available online at http://dl.acm.org/citation.cfm?id=1979436&CFID=591545870&CFTOKEN=33599456.

Figures 4.3, 4.4, 4.7, and 4.8 are from an article by the authors published in *Journal of Computer Supported Cooperative Work*, 2014 (copyright Springer), available online at http://link.springer.com/article/10.1007/s10606-014-9203-4.

Figure 4.9 reproduced with kind permission from Mithun Jacob and Juan Wachs.

CHAPTER 1

Introduction

Understanding the posture and motion of the human body has been an enduring interest and concern within healthcare settings. For well over a century now, gait analysis work has sought to systematically characterize and quantify different aspects of the body in locomotion (e.g., Whittle, 1996a, 1996b). In particular, these efforts have looked to articulate the patterns and parameters of a normal healthy gait as well as the deviations from a normal gait that are characteristic of particular forms of clinical movement disorders from neurological conditions through to musculoskeletal conditions. Augmenting the observational methods of experts, our understanding of the moving body has been furthered through evolving forms of technological development and instrumentation. The early photographic techniques pioneered by Eadweard Muybridge (1893) and Étienne-Jules Marey (1890) gave rise to important advances in the science and understanding gait analysis over and above what was possible with unaided observational approaches. Following this came the widespread introduction of the video camera, which again opened more sophisticated ways in which characteristics of human movement could be systematically measured, analyzed, and understood. In more recent decades, the field has witnessed even more dramatic technological developments with the emergence of a wide variety of sophisticated computational sensing technologies and techniques. Accelerometers, gyroscopes, goniometers, smart fabrics, force plates, and electro-magnetic sensors, as well as a marker-based and marker-free computer-vision techniques, all offer new ways in which different features of bodily posture and locomotion can be tracked, measured, and interpreted.

These developments have been significant for a number of reasons. In part, this evolving instrumentation offers an increasingly diverse set of ways that features of body movement can be sensed, measured, classified and represented. These may include the spatio-temporal properties of individual joints and limbs (e.g., position, acceleration, rotation), spatio-temporal measures of particular actions (such as stride length, gestures), tracking and representations of complete action trajectories, kinetic forces, and outcome measures such as achievement of particular motion end goals (e.g., successfully reaching for an object). Their significance lies also in efficiencies that they can bring to the analysis of body motion data, reducing the cost and time of data capture and making such data accessible to richer more sophisticated mathematical analysis. This has led to accelerated developments in the fundamental understanding of normal human motion as well as in our understanding and assessment of the movement abnormalities and symptomatic movement patterns (e.g., tremor, freezing) manifest in clinical conditions such as cerebral palsy (CP), Parkinson's disease, multiple sclerosis (MS), chronic pain, and stroke (e.g., Dobson et al., 2007; Hass et al., 2005;

Rodda et al., 2004; Salarian et al., 2004; Morris et al., 1996; Socie and Sosnoff, 2013; Ferrarin et al., 2014). This understanding is important both for diagnostic assessments and treatment planning but also to enable us to understand the impact of particular interventions on the disease progression.

The above significances relate well to the core aims of traditional gait analysis work within the healthcare domain in the sense that they are concerned primarily with notions of measuring and analyzing the body with a view to achieving some form of clinical interpretation. But there is additional significance in this digital instrumentation of gait and motion analysis that can be found in the interactive opportunities that have opened up. The digital sensing and tracking of the body in motion now presents opportunities for computational behaviors and responses to be made contingent upon digitally represented patterns of human motion. In this respect the moving body is no longer just a quantifiable object of analysis and classification in itself as it might be conceived within more traditional orientations to gait analysis and clinical assessment. Rather, the body can be transformed in profound ways by the ability to digitally infer and dynamically respond to its patterns of motion. This further opens up the ways in which we need to understand the properties of these body-tracking technologies in healthcare because of the system behaviors that can now be attached to their output. Here then, it is not just what aspects of the body in motion can be tracked and quantified but also what the system can action on the basis of this data. It is the data in application that opens a broader scope of enquiry in relation to these technologies and for which there is a larger set of socio-technical concerns at play.

If we think about the use of body-tracking technology within rehabilitation scenarios for example, the more traditional focus of gait analysis has been with the assessment of particular movement disabilities displayed by a particular individual. Such assessment then forms the basis of planning appropriate and bespoke treatment regimes or evaluating the success of particular clinical intervention on movement disability severity (e.g., Kuan et al., 1999; Salarian et al., 2004; Stokic et al., 2009; Lopez-Meyer et al. 2011). But with the widening scope enabled by the interactive capabilities of these body-tracking technologies, we are now seeing growing interest in rehabilitation scenarios that extend beyond the analytic offerings of earlier forms of gait analysis. So, for example, rehabilitative treatment interventions have been developed that provide various forms of feedback to a patient based on particular forms of clinically relevant bodily motion with a view to facilitating the patient's understanding and awareness of their movement (e.g., Singh et al., 2014). Similarly, numerous examples of interactive serious games and exercise programs have been explored to exploit motion-based tracking control as a means to encourage particular forms of body movement in the context of a patient's movement rehabilitation (Alankus et al. 2010; Geurts et al., 2011; Jaume-i-Capó, 2014).

Other examples opened up by the possibilities for actively responding to the tracked body include everyday *activity monitoring*. Here, rather than using the body-tracking technology for clinical assessment, the opportunities are in ongoing sensing of movement in everyday life and inferring particular forms of behavioral action states from the sensor data. An example here would be fall monitoring in older adults where body worn accelerometers can be used to infer when the characteristic patterns of a fall have occurred. On detection of the characteristic data pattern various things are consequently triggered such as sending a notification to a caregiver or clinician. The impact of this on the everyday practices and lives of the patients in question extends beyond the more traditional realms of gait analysis and this need to be understood in new ways. Other clinically significant patterns of movement disorder have the potential to be inferred in the context of everyday living. For example, the gait initiation and freeze states associated within Parkinson's disease may be inferred in ways that could allow the triggering of new forms of interactive clinical intervention.

But such digital instrumentation of body movement analysis also offers opportunities in healthcare contexts that extend even further outside the bounds of more traditional gait analysis. More specifically here, we can shift our attention from patient-centric application of these technologies (where the form and motion of the patient's body is of central concern) to new forms of clinician-centric applications. Here it is the body of the clinician, not the patient, which can be also considered as a focal point for body-tracking technologies to enable new forms of body and motion-based interactive possibilities in healthcare scenarios. Of particular interest here has been the context of the operating room where body tracking of the clinician offers the potential to overcome the specific constraints of these environments imposed on more traditional interaction techniques. By tracking the body movements of the clinician as opposed to the patient in these contexts, control of surgical equipment, for example, can be achieved touchlessly with these technologies without compromising sterility.

With these developments in technology, then, we have greater number of ways that body movement can be captured, measured, and interpreted, but also a much broader scope of potential application areas for the healthcare domain. Understanding these technologies and their clinical significance raises a diverse set of concerns. On the one hand, one might look to understand these technologies in the terms of particular characteristics of movement that they capture and measure and the performance parameters of the technologies relating to these—for example their accuracy, precision, resolution and reliability with respect to different types of movement representation. While this provides a useful first basis for thinking about these technologies, it is quickly apparent that such characterizations are not sufficient in and of themselves. One cannot simply consider the performance attributes of these technologies in terms of their relative distance from an ideal way of representing the body in the richest and most accurate way possible. So while greater motion tracking performance in terms of resolution, reliability, and accuracy may open up new opportunities they are not end goals in themselves. The more pertinent concern here is what needs to be measured and how well for a particular clinical scenario at hand—be that understanding the movement patterns of a clinical disorder or sensing movement necessary to achieve particular interactive outcomes. Each of these contexts, uses, and clinical conditions may demand more or less of particular forms of body tracking and may be robust in different ways to the respective limitations and characteristics of these technologies. One of the goals of the book is to highlight some of the complex interdependencies between the technical readings of the body's motion and clinical perspectives. The aim here is to allow a greater reflection on some of the assumptions bound up in body-tracking technologies and algorithms with a view to understanding their clinical significance.

1.1   ENABLING TECHNOLOGIES

To help us further these discussions, we first take a look at some of the key technological approaches for measuring and understanding different aspects of the body in motion. One of the key approaches concerns *Kinematic* characterizations of the moving body, which articulate geometrical features of body motion such as limb/joint position, orientation, displacement, velocity, and acceleration. These geometrical features of motion may be articulated in two and three dimensions depending on the requirements of the scenario and may apply to specific limbs and joints or the moving body as a whole.

1.1.1   CAMERA-BASED SYSTEMS

Many kinematic systems for tracking and measuring the moving body rely on some form of *camera-based* set-up. Simple systems can be based around single camera set-ups though these can present certain limitations in terms of orientation flexibility and in terms of opportunities for tracking in three dimensions. Many of the more sophisticated commercial set-ups used in high end gait laboratories though rely on multi camera set-ups to enable orientation flexibility and full 3D capture of limb and joint movement. While there are variants in the different systems, these high-end systems generally rely on some form of marker tracking. Here different types of markers are attached to the joints and limbs in standardized configurations such as the Helen Hayes (Vaughan et al., 1992) and CAST (Cappozzo et al., 1995) marker sets. These may be passive systems in which retro-reflective markers reflect infrared or visible light back to the sensing camera. Alternatively, some systems use active marker systems such as opto-electronics that employ light emitting diodes (LEDs) as markers. The advantage of the active systems is that they can more reliably identify and track the individual markers from one frame to the next leading to a much less noisy and stable tracking of the markers.

Generally speaking, most of the commercial camera based systems offer good accuracy in terms of position data, somewhere in the region of 1–3 mm in all dimensions. Measurements of acceleration and velocity can suffer when there is even a small amount of noise in the original data, though these can often be compensated for through the application of appropriate filtering of the data if this is not detrimental to a particular clinical scenario. Other challenges arise from the requirements for accurate positioning of the markers correctly on the body that can require a certain level of expertise on the part of the operator. Estimation techniques for this positioning are subject to certain errors while there are also sources of error arising from the movement of markers relative to their intended position with respect to the bone.

Such set-ups, while producing good quality data, can be cumbersome and expensive to set-up, requiring large amounts of dedicated space that is difficult to reconfigure and move. For many application possibilities that may be less sensitive to fine grained accuracy requirements, these may be well beyond the scope of what is necessary. An intriguing shift in recent years has been the rise of much cheaper camera-based alternatives derived from commercial camera-based computer gaming systems such as, most recently, Sony's PlayStation camera and Microsoft's Kinect. The Kinect system in particular has received particular interest in the clinical communities that has begun to assess its feasibility for a range of rehabilitation, assessment, monitoring, and control scenarios (e.g., Gabel et al., 2012; Stone and Skubic, 2011a, 2011b). Significant here is that it is offers a much more affordable alternative to the high-end gait laboratory tracking set-ups. Similarly, while it is fixed during actual use, it is relatively portable making it suitable for easy deployment in a wider range of environments than more high end gait lab set-ups (e.g., home). The Kinect system combines VGA and depth sensing cameras and, as well as producing a raw depth image of any scene and body sampled, software is used to infer and extract a virtual skeleton model of the scanned body to be tracked in real time (Shotton et al., 2011). The inferred skeleton in the Kinect SDK comprises 20 points based on rough estimations of the center of the major joints in the human body meaning that tracking is not dependent upon the placement of external markers on the body—the set-up effort and costs entailed by marker-based set-ups. Such inferred estimates of the major joints though do come with certain compromises in terms of the precision. Such precision may be suitable for coarse-grained gestural interaction and simple quantifications of motion parameters (Bonnechère et al., 2014). The inferred skeleton of the Kinect is also a planar representation so it does not constitute a full 3D representation of the human body which again creates certain constraints in the ways such technology might be deployed. This also has implications for the organization of body movement in the sense that the tracked body assumes the body is facing toward the camera. In its out-of-the-box form, it may be less suited to visualizing and understanding of the complex motion patterns associated related to the pathologies associated with cerebral palsy, stroke, chronic pain, etc. Such constraints though may be overcome by supplementing the inferences with anatomically correct motion modeling data from validated biomechanical studies of complex movements (Bonnechère, et al., 2013). The important point highlighted by Bonnechère is that the characteristics of any enabling technology for motion tracking is in part related to the fundamental sensing capabilities, and in part related to the assumptions and inferences of the associated computing algorithms. Both of these need to be understood in relation to the application opportunities of the particular body-tracking technology.

1.1.2   BODY WORN SENSORS

A significant alternative to vision-based approaches to kinematic motion tracking in healthcare can be found in the use of body worn sensors (e.g., Morris, 1973). Such sensor-based approaches are of appeal in the sense that they are relatively inexpensive and impose fewer constraints on the area of movement. There are a number of commonly used sensors such as accelerometers, gyroscopes, and goniometers that are either held, worn, or attached to particular parts of the body such as the leg, foot, arm, waist, etc. (e.g., Crea et al., 2014). *Accelerometers* are a form of inertial sensor that measures acceleration along a particular axis of movement and from which velocity and positional data can be derived—although there are generally accepted problems of "drift" with accelerometers since errors in measurement accumulate from point to point and build up over time. There are three common variants of accelerometer, piezoelectric, piezoresistive, and capacitive with capacitive generally having higher levels of sensitivity and resolution (Wong et al., 2007). In order to enable the sensing and measurement of rotational movements, accelerometers are often combines with *gyroscopes*, a form of inertial sensor that measures angular velocity. Again, when attached to particular parts of the body, this angular motion velocity can be used to measure and classify key features of body movement (e.g., Ayrulu-Erdem and Barshan, 2011; Tuncel et al., 2009). These can also be complemented with data from *magnetometers*, which essentially allow position and orientation of the sensor relative to the magnetic field (Graham et al., 2004). While magnetometers are relatively poor for accurate assessment of fast movements they do not suffer from the same levels of drift over time and thus can compensate for some of the drawbacks of the other two types of sensor in helping to determine accurate orientation and position. Advances in microelectronics now means that these sensors scan be combined in a single unit. As with some of the marker-based camera approaches, wearable sensors can suffer some similar challenges in terms of the accuracy with which they can be placed relative to corresponding body parts.

A further kinematic characteristic of movement concerns the relative motion between different body segments. The continuous measurement of the angle of a joint between two body segments can be achieved using various forms of *electrogoniometers*, based on strain gauges, potentiometers, and mechanical flexible techniques. While a single goniometer will measure angles in one plane it is possible to mount multiple goniometers in different planes for multiaxial tracking requirements.

With advances in material sciences, sensing fabrics (e.g., de Rossi et al., 2001; Sawhney et al., 2006; Scilingo et al., 2003; Mazzoldi et al., 2002) also offer an intriguing way to track measure characteristics of body posture and motion. Here fabrics are imbued with particular sensing attributes by depositing different forms of polymers and materials onto or into the fabric. The deposited materials can have different resistive and capacitive qualities that dynamically change as the fabric is stretched. While these systems vary in their responsiveness they are important in the ways that they begin to highlight some of the more practical elements of sensing activities in particular for adoption in more every day healthcare scenarios. The fabrics offer a lightweight and ergonomically comfortable way of getting sensing on the body and thus may have benefit for body-tracking applications that take place in everyday contexts over longer periods of time.

1.1.3   FORCE AND PRESSURE-BASED SYSTEMS

As well as articulating the basic patterns of motion posture in *kinematic* terms, there are also important elements of body tracking that can be grounded in more *kinetic* terms. Such kinetic characterizations are concerned with how movement manifests in terms of force and pressure. Force platforms, for example, are commonplace within gait labs. Such platforms consist of a rigid upper surface layer. Beneath this layer is a set of transducers positioned appropriately to measure small displacements of the upper surface in three axes. What is of interest in such systems is while the foot is the point of contact with the system, the recorded forces actually pertain to the motion of the body as a whole, representing the acceleration of the center of gravity of the whole body. From such analysis of the end points of motion we can determine something more holistic about the ongoing posture, sway and balance of the body. This information can be important for diagnostic purposes, rehabilitation, and interactional control. Interestingly, derivatives of this technology have found their way into commercial gaming systems such as the Wii Balance Board. As with the Kinect system discussed earlier, the significance of this lies in making these tracking and measuring capabilities available at a much lower cost and in a way that is accessible in a wider range of environments outside of high end gait laboratories, such as the home. Other force sensors may be worn on the body and, in particular, embedded into footwear (e.g., Faivre et al., 2004) creating opportunities for kinetic sensing of movement within an even broader range of everyday circumstances.

While these sensing technologies form the fundamental basis for tracking the body, the captured data is actually only the starting point. In order to make sense out of the captured data, additional analytic techniques need to be applied. For example, various forms of algorithmic filters may be applied in to compensate for particular features and shortcomings of particular sensing technologies in relation to certain forms of body motion. Such filters discard or transform selected unwanted elements of the data while maintaining other relevant parts.

1.2   BODY TRACKING IN CONTEXT

As we suggested above, presenting an overview of the enabling technologies in this way allows us to get an initial sense of the possibilities. But the driving concern in such treatment remains with the centrality of body measurement and analysis as the end in itself. That is, what can be reliably sensed and inferred with a view to how these tracked and measured actions might align with symptomatic properties and clinical indicators or with forms of interaction techniques appropriate to particular clinical settings. In many senses this has been an appropriate focus for the field of gait analysis as it has been dominated by its research efforts. But in relatively recent years, these research efforts of gait analysis and body tracking have started to find more significant application in routine clinical and every day practice. For example, we see it commonly in the clinical management of cerebral palsy where is may be requested as part of orthopaedic surgical planning processes as well as post surgical analysis. We also see its use being more routinely requested in the assessment of an individual's motor dysfunction post stroke. This shift is all set to increase as research matures but also as various forms of body-tracking sensors pervade ever more areas of our everyday lives beyond the bounds of traditional clinical contexts. One only needs to consider the ubiquity of smart phone technology and the recent growth in smart wearable health bands to get a sense of how widespread body-tracking technologies such as accelerometers and gyroscopes have become. In addition, with the ability of digital body-tracking systems to functionally respond in real time to the sensed data streams, we will see wider applications and uses being situated in everyday clinical and life practices.

The shift from research to more routine everyday practice has important consequences for how we need to consider these various technologies in healthcare contexts. While the nature and form of bodily movement and its inferred representation remains a central concern, there is now an additional imperative to consider how the material form of these tracking and assessment opportunities can be situated within the larger systems of clinical workflows, environmental contexts, social and collaborative practices and the everyday contexts of routine living. Much of the concerns of these shifts into practice remain relatively unexplored in gait research but, as we will discuss later in the book, there are profound and mutually interdependent ties between the fundamental characteristics and assumptions of these technologies and the configurations of everyday collaborative practices with them. The field is now at a point where greater attention to these concerns is becoming of critical importance to the ways that these technologies can be appropriated.

Such a practice-centric perspective points to a number of key socio-technical considerations in the understanding of these body-tracking technologies. In the first instance there is an important consideration about how we conceptualize data and measurement of the body derived from these technologies. Rather than conceiving the data and measures derived from these technologies purely as self-explicating representations of the objective reality of the moving body, it is important to consider how they are embedded in practice and how the actions on this data render them meaningful in particular contexts. Such arguments are derived from key thinkers in *Social Studies of Science* and technology (e.g., Lynch, 1985, 1990a, 1990b; Goodwin, 1994, 2000). But increasingly these arguments are finding significance in more healthcare contexts and design oriented disciplines in ways that not only situates them closer to our domain of concern but also offers the basis for reasoning about particular technological approaches in terms of their practical implications (e.g., Hartswood et al., 2003; Mentis and Taylor, 2013; O'Hara et al., 2014a). In their arguments about medical imaging technologies and the visual representation of the body, Mentis and Taylor (2013) highlight the importance of moving away from simply "privileging image fidelity and detail" toward a greater privileging of the "active use of images." Likewise, as Lynch (1985) argues, we need to consider data in terms of a *"representational adequacy"* that is not fixed but dependent upon a *"coherence of actions established in the social environs"* of a setting (p. 60).

The "privileging [of] image fidelity and detail" arguments in Mentis and Taylor (2013) relate nicely to the *"centrality of body measurement"* perspectives highlighted above in relation to the currently narrow ways of talking about body-tracking technologies. In part, this points us to the need to attend further to the *in situ* and interactional organization of any body-tracking data that can be collected through this varied array of sensing technologies; the ways it can be visualized, constructed, attended to, pointed at, oriented to, manipulated, and discussed in the context of particular clinical activity. That is, what can be done by the human actors in clinical setting that are, in Lynch's (1985) terms, *disciplining* the data? This practical achievement of how such data is given meaning in practice is arguably one strand of work that is under articulated within the more traditional realms of gait analysis but is starting to gather attention within fields such as Human-Computer Interaction.

What this points to, then, is a greater need to think about these technologies in terms of the opportunities they create for reconfiguring routine practices which entails detailed understanding of the socio-technical context in which they are embedded. As a simple high level illustration of these issues let us consider how much of the early gait assessment and body-tracking work has been reliant upon specialized and dedicated gait assessment labs. While these labs are optimal from the perspective of enabling the very highest quality of gait and body movement data to be collected, such a privileging of high quality quantification comes with particular consequence in terms of practice. So, for example, these set-ups are enormously costly from a financial perspective. In addition, they are complex and time consuming to set-up, and demand a high levels of expertise to run. Such characteristics ultimately constrain and shape the ways these technologies can be appropriated in particular clinical trajectories, workflows, and practice. If we take the simple issue of cost, this factor alone is prohibitive to a more ubiquitous deployment of these technologies in other forms of clinical setting or indeed in the home or on the move. Being a high cost resource means they are also a more scarce and limited resource. The consequence of this is that any quantification of patients' posture and motion requires them to travel to a particular place at a particular narrow time windows. Already then we can start to see how this limits opportunities for where and how often such tracking can be carried out preventing more frequent and ongoing tracking over the course of time and particular disease/condition progression. What this begins to do is turn our attentions to a different set of factors and trade-offs that are important to consider in relation to these technologies over and above the centrality of body measurement concern. So, for example, one might consider trading representational precision and accuracy for lower-cost sensing technology because of the potential to open up usage of such technologies outside specialist centers—e.g., in health centers, and in the home.

The significance of this may go beyond simply making the availability and accessibility of these technologies more widespread. Indeed, in many practical applications of these technologies, key elements of their success arise by virtue of the practices being situated in a particular social setting. This may be the case for example, with the issue of rehabilitative gaming using body-tracking technologies. Here while body-tracking technologies and games can be designed to promote particular forms of rehabilitative movements, this in itself may be pointless if the patient does not have the appropriate motivation and incentive to perform the movements in the first place. As we discuss later, such motivational requirements may be achieved by situating the technologie,s in particular social contexts such as a care home where a game-based rehabilitation regime can be performed in the company of other residents. Such an opportunity only really becomes plausible when the sensing technology is low cost and portable enough to be deployed in these kinds of social contexts. If a sensor, too, is something that can be worn comfortably for hours and even days, this opens up the possibility for the continuous or more frequent capture of body movement data in the context of everyday living relative to the rather artificial setting of a gait lab. This is not simply a question of ecological validity of gait data but more of an important realization of the ways in which the motor capabilities, performance, and dysfunctions may be manifest, constrained, and managed in very different ways depending on the social, physical, and environmental circumstances.

Consideration of the tracked body as a *social body* in the presence of other actors is an important shift that reveals many different implications in relation to particular forms of body tracking. For example, there may be issues of social acceptability of particular forms of tracking technologies that may be conspicuous in the presence of others. Irrespective of how well certain technologies may operate in principle, if they lead to various forms of social discomfort, this will ultimately shape their patterns of use in practice. Of further interest too is how various forms of body-tracking technologies impose particular influence not just on an individual being tracked but on the movements and actions of other people in the vicinity of the tracked body. For example, the algorithms in computer-vision tracking systems may demand that the moving bodies remain visibly separated or may demand particular orientations to a camera rather than a collaborating actor. In this respect, the tracking characteristics can affect the interpersonal configurations and relations of the multiple actors making up different healthcare contexts as the actors work together to make the sensing systems work. This could include, among many others, the relationship between doctor and patient during assessment, or the relationship between collaborating clinicians during a surgical procedure. If we take the doctor patient relationship as an example here we will discuss later how these considerations become a very real issue in the organization of assessments of patients with movement disorders. With such patients often finding difficulty performing certain movements unaided, the clinician may offer support bringing the body of the clinician together with the body of the patient. Treating these connected bodies as distinct entities can pose a particular challenge for computer-vision-based body-tracking systems that rely on the visual separation of the bodies.

Here we see illustrated the intertwining of social and technical circumstances in ways that bears on our understanding of these technologies in context.

There are many such social, clinical, and technical relations pertaining to these technologies that may play out, in particular ways, across a range of different healthcare contexts and across a range of different stakeholders. Critical in the future development and successful adoption of these technologies, by clinicians and patients alike, is a greater emphasis on understanding the socio-technical contexts and implications of these various approaches to body-tracking technologies in different healthcare scenarios. One of the aims of this book is to offer the reader various examples of this kind of discussion and illustrate some key ways in which social contexts of use relate to the properties and assumptions bound up in particular choices of body-tracking technology. In addition, we can see how certain technology characteristics might lead to the collaborative practices of these healthcare scenarios to be configured in particular ways. The intent here is not to offer a complete and comprehensive account of all scenarios of use but, rather, through the examples, offer critical insights and more importantly a perspective that can be brought to bear on a broader set of technologies and scenarios that relate to this exciting field.

1.3   OVERVIEW

In presenting our discussion of these technologies, this book explores three core application areas in healthcare. Each of these application areas offer us different ways of thinking about the body in motion as something that can be tracked and understood toward different ends. Here we are able to articulate how specific demands of the application area and the settings in which they are practiced have particular implications for how a common set of body-tracking technologies can be critically understood. In the first instance we look at the issue of body tracking in the context of clinical assessment. In many ways this has been at the heart of more traditional gait analysis research and practice efforts to date. While there have been a plethora of feasibility studies in the area that point to many ways that body tracking and analysis can be used to infer particular neurological disease states, the practical accomplishment of these assessments remains a richly intriguing challenge for a variety of reasons. Using assessment of multiple sclerosis as a core example, the chapter sets out to reveal the practical challenges faced in using ubiquitous sensing technologies for assessing movement disorder in actual practice.

The second area shifts emphasis to consider these technologies in the context of rehabilitation; more specifically scenarios of self directed rehabilitation and care. Here two different areas of rehabilitative care are considered. The first of these concerns the treatment of chronic pain while the second looks at reducing fall risk in older adults through the bespoke forms of balance training. Of interest in the treatment of chronic pain scenario is the realization that any movement-based disabilities experienced are not simply bound up in physical concerns. Rather, their manifestation is bound up also with psychological and affective states. In this regard, body-tracking technologies in these scenarios are not simply about how they can encourage particular forms of physical movement but also how systems might be designed to recognize or influence psychological and affective factors. Of significance here are the ways in which the body-tracking approach moves beyond analysis and assessment to consider the role of interactive outcomes tied to specific bodily action. Interactive and motivational concerns are apparent too in the context of the balance training and rehabilitation as well as the ways in which particular forms of balance related movement can be encouraged. Here motivation in rehabilitation is treated as a multi-faceted concern that has implications for how we conceive the role of body-tracking technologies in these scenarios. Rather than just tracking body position and control to understand balance performance, it looks at how balance indicators influence events within a game, which are manipulated to get the patient to move the body in progressively more difficult ways. In this respect there are motivational factors intrinsic to the game and task. But the work also highlights some of the more social aspects of motivation in rehabilitation and how this arises because the form of the technology enables it to be embedded in a particular social setting.

The third theme in this book then takes a rather different turn. Building on some of the interactive opportunities of these technologies, the theme considers how the movement of the body can be used to develop new forms of interaction of relevance to healthcare. More specifically, the significant turn here is to focus on the body of the clinician as also something to be tracked and explore the implications of this in the particular setting of the operating room. Of importance here are some of the unique demands of these settings that lend significance to opportunities for touchless interaction and hands free interaction. In addition, the discussion here is used to highlights ways in which certain task demands as well as the social and collaborative organization of the setting relate to particular characteristics of the motion tracking and inference technologies.

Our aim in this book is not to offer a complete account of these technologies within particular clinical applications and contexts. Rather, through a range of examples, our aim is to introduce a richer set of concerns and perspectives that are important to consider when thinking about these body-tracking technologies in healthcare contexts. The intention is for these examples to be used as a starting point for the reader's own reflections on how new emerging forms of body-tracking technology may come to be used and appropriated in a range of new application areas and settings.

CHAPTER 2

Clinical Assessment of Motor Disability

2.1 INTRODUCTION

Neurological conditions, such as multiple sclerosis (MS), cerebral palsy (CP), Parkinson's disease, and stroke cause deterioration in a patient's motor abilities. Such deterioration can impact balance, walking ability, and the dexterity needed for everyday tasks, such as eating and drinking. While there may be significant variation in motor deterioration exhibited by patients, condition-specific patterns of movement and gait are manifest (e.g., tremors, freezing and gait initiation problems exhibited in Parkinson's disease). A large number of clinical interventions and therapies have, and continue to be, developed to relieve symptoms that challenge daily living, such as ataxia (swaying of the body) or dyskinesia (involuntary muscle movements). Being able to assess, characterize, and/or measure particular aspects of motor performance is a key element for successful clinical intervention.

Such assessment may be for the purposes of classifying types, stages or severity of a patient's condition. It may also be for the quantification of particular movement traits with a view to determining the specific details of an intervention bespoke to a patient. For example, one of the more successful areas of gait analysis in real clinical scenarios is in surgical planning for children with CP. Analysis of gait and movement parameters here is used to identify the specific combinations of orthopedic procedures required by a CP patient. But it can also be useful in the planning of other relevant therapies. Sticking with the example of CP, analysis and measurement of gait patterns can help identify which muscles in a CP patient might be appropriate sites for botulinum injections, which suppress the activity of neuromuscular junctions and therefore prevent activity of the problematic muscle. Likewise, understanding particular manifestations of dysfunction can be used in the design of certain physiotherapy regimes or the specification of bespoke orthotics for a patient. As well as determining the forms that clinical interventions might take, clinical assessments of motor performance are an essential component in understanding the impact that interventions may be having. This understanding of disease progression applies both to individual patients for whom the effects of an existing treatment need to be determined and to clinical populations at large for which the effects of new treatments can be assessed in the form of clinical trials. It would be standard clinical practice, for example, to perform a follow-up gait analysis on a CP patient one or two years after having surgery and standard practice to perform robust clinical trials of new drug interventions.

High-quality clinical gait analysis techniques available in gait laboratories do not always translate well to other sites of clinical practice, both in terms of their practicalities and in terms of what they offer in terms of motor assessment possibilities (Simon, 2004). For these reasons, in many areas of clinical practice, motor ability continues to be measured through the use of standardized rating scales based on an expert examination by a clinician. These classification scales include the Expanded Disability Status Scale (EDSS) (Kurtzke, 1983), the Gross Motor Function Classification System (Palisano, et al., 1997, 2008), and the Unified Parkinson's Disease Rating Scale (UPDRS) (Fahn et al., 1987). When using these scales within a clinical assessment context, patients are asked to do a series of specified functional movements, such as moving keys from one hand to another, walking for 10 m, or touching the finger to the nose. Exercises are then graded for specific abnormalities on an ordinal scale. In most tests, scores on specific exercises are then combined and summarized into system scores and/or a global score. The exercise, system, and global scores may be used to characterize patients and judge specific treatments.

Despite careful design, standardized rating scales have inherent issues of reliability and sensitivity (Hobart et al., 2007). The issues are not so much with the clinician's professional vision and what they can ascertain on the basis of various movement characteristics. Rather, the apparatus for classifying and articulating these assessments in a meaningfully quantifiable way are arguably somewhat blunt. In addition, such classifications are not made independently of any action and treatment plans that may follow on from particular ratings and classifications. Indeed such classifications are often made explicitly with the intention to trigger particular courses of intervention rather than simply being an accurate articulation of a disability (cf. Hartswood et al., 2003). As a consequence of this, agreement between clinicians (inter-) and the same clinician over time (inter- and intra- rater reliability respectively) is usually low. When we consider these measures in the context of tracking disease progression over time these issues of inter- and intra-rater reliability become problematic. In order to compensate for this and increase agreement across clinicians and across time, the granularity of the scale is commonly reduced (i.e., a scale of 0–7 is reduced to 0–4), which, in consequence, makes the scale less sensitive to changes in disease state. As examinations take about 1 h of a neurologist's time, they are also very expensive. As such, while these scales help structure an examination and provide at least an initial quantitative measure to support a clinician's overall view of the patient, they are not reliable for tracking disease state.

The challenges of using these standardized rating scales for assessments that can reliably track disease state across patients over time can lead to various difficulties in assessment practices. In particular, it can make it difficult to ascertain the impact particular interventions are having on a patient's motor dysfunctions and whether alternative courses of action might be more appropriate. Similarly, when developing new forms of drug treatments or other novel interventions, such lack of reliability and classification granularity can hinder the ability to convincingly demonstrate effects in the context of a clinical trial. From a pragmatic perspective here, the uncertainty levels associated with these scales means that clinical trials often require larger numbers of clinical participants in order to demonstrate effects with acceptable levels of power. This makes it problematic to measure the efficacy of new drug treatments during a clinical trial with a reasonable and manageable number of patients (e.g., Cohen et al., 2012). One of the key motivations for more sensitive and reliable motor ability assessment over time is to make the evaluation of treatment and drug trials shorter and cost effective. So, for example, by reducing the uncertainty associated with a particular assessment measure two-fold in MS trial, we would enable approximately a four-fold smaller sample size while preserving the power of the trial. A further key motivation for more sensitive and reliable assessment is that it opens up opportunities for new reactive treatments that change with disease state (e.g., Gladstone et al., 2002; Leddy et al., 2011).

This leads to somewhat of a quandary for the organization of clinical practice. On the one hand, the gait analysis laboratories offer opportunities for more precise, accurate, and reliable motion tracking. But their practical requirements and cost place huge restrictions on the their more pervasive deployment across different potential sites of clinical intervention. On the other hand, the various movement classification scales may be limited in their precision and reliability but do have the benefit of being more widely deployable across a range of different healthcare contexts. The factors at the heart of this quandary have motivated keen interest in recent developments in sensor technology, robust wireless data transfer, and the miniaturization of electronics. Such developments utilizing low-cost consumer-grade sensors (e.g., depth sensing cameras, triaxle accelerometers and gyroscopes) have the potential to produce more cost-effective, less contextually demanding and easy-to-use systems. One prominent aim of developing ubiquitous sensing applications for the treatment of neurological conditions is to provide a consistent, granular, quantified measure of motor ability—a technically viable alternative to the currently used subjective standardized movement assessment rating scales that could potentially overcome inherent issues of reliability and sensitivity in these scales (Hobart et al., 2007). Moreover, automated sensor assessment could reduce the cost of measurement. The resulting measure would be useful for tracking disease progression and supporting clinical trials in neurological diseases.

Researchers and engineers have quickly adopted these sensors to develop innovative applications for clinical intervention for neurological patients. The demands of movement measurement for clinical assessment has made application development more challenging but a number of research efforts aim to validate the measurements produced by these sensors against benchmark marker-based motion capture systems. With depth-sensing cameras, for example, the evidence so far suggests that Kinect is appropriate for functional ability assessment of motion in healthy people (Bonnechère, et al., 2014). Others studies have focused on specific characteristics of clinical conditions, such as: postural control (Clark et al., 2012), upper extremity function in dystrophinopathy (Lowes et al., 2013), static foot posture (Mentiplay et al., 2013), and continuous in-home gait analysis (Stone and Skubic, 2013). There are initial explorations of the use of depth sensing technologies for carrying out gait analysis in MS patients (e.g., Behrens et al., 2014; Gabel at al., 2012; Stone and Skubic, 2013). Similarly body-worn motion sensors have been demonstrated to highlight mobility differences between healthy volunteers and patients with early-stage MS better than traditional time tests (Spain et al., 2012), while accelerometry has also been used to measure physical activity, gait and walking mobility in MS patients (Weikert et al., 2010). Most recently, an accelerometer built into an iPad has been used for gait and balance analysis (Rudick et al., 2014). In patients with Parkinson's disease, wearable and mobile device sensors (accelerometers, gyroscopes and magnetometers) have demonstrated feasibility in quantifying various motor symptoms such as tremor, rigidity, bradykinesia, freezing gait, dyskinesia, and postural instability (e.g., Hoff et al., 2001a, 2001b; Thielgen et al., 2004; Kim et al., 2010; El-Gohary et al., 2014; Mazilu et al., 2014; Baston et al., 2014; Horak and Mancini, 2013; Weiss et al., 2011).

These initial explorations point to a promising potential for these new forms of low-cost ubiquitous sensor. As well as offering some form of middle ground resolution to the aforementioned quandary, these technologies promise even more dramatic changes in terms of the ways that clinical assessment and interventions for patients with motor dysfunction can be configured in practice. For example, ubiquitous sensing technologies can enable movement assessment in a wider range of clinical environments outside of specialist gait laboratories and even in nonclinical settings to assess movement as it occurs in naturalistic settings and contexts, unencumbered by the stress of visiting a clinic (e.g., Kiani et al., 1997). This could enable the measurement of Activities of Daily Living (ADL), such as drinking from a cup, which demonstrates the impact of a disease on patient's quality of life. Body tracking and motion analysis could potentially be performed more frequently and over longer periods of continuous time, with a home measurement system. By tracking and assessing motor function more frequently, ubiquitous body sensing and tracking technologies could open up the development of more reactive treatment regimens, which change with disease state. Systems that support symptom management, such as freezing gait in Parkinson's, can provide immediate relief (Takač, 2013). Rehabilitative programs can also be built on sensed measures, often with the assessment and rehabilitation coming together into one system (e.g., Young et al., 2011).

While this early work exploring the feasibility of these systems for tracking and measuring various forms of symptomatic movements within clinical conditions, the translation from this to active use in clinical practice is less trivial than it may initially seem (e.g., Morrison et al., 2016). Such a translation pulls into focus a different set of characteristics in relation to these ubiquitous sensing technologies that pertain not just to their demonstrable ability to track and measure bodily movement but rather to the ways that assessment workflows and practices are and will be (re)configured. Careful thought is needed to understand how these technologies might be integrated into current workflows.

Relevant lessons can be drawn from O'Hara et al. (2014a, 2014b), which sought to understand the use of these motion-tracking technologies (more specifically depth sensing) in the operating room. The use of the depth sensing in this instance was not for assessment purposes but to enable clinicians to interact with medical images in a sterile manner using touchless gestural interaction techniques. While it was not the first system to demonstrate the feasibility of depth sensing for gestural interaction, the point of the work was to move beyond those questions of gesture control feasibility. Its aim was to begin articulating how system design specifics related to features of clinical practice in these settings and how clinical practices were reconfigured in response to characteristics of the depth sensing system. As an illustrative example, the research, discussed how *line-of-sight* requirements of the depth sensor gesture technology impacted on access and control of the system by different clinicians in the OR, how background members of the clinical team were inadvertently and unknowingly implicated in the interaction, how it affected the spatial configuration of actors in relation to the sensor and particular demands of the machine learning algorithms and displays. The key point of the work was to highlight a more intimate relationship between fundamental elements of the technology and the features of the clinical practice that broadens the ways that we conceive of the design of these systems. Drawing on these ideas, we explore more closely the alignment between characteristics of a ubiquitous sensing and the clinical practices of assessment.

To help us in relating these technologies to the practical accomplishment of clinical assessment, we will use some reflections on our own experiences in developing a motion tracking-based computer system for use in the clinical assessment of the neurological condition of MS (e.g., Morrison et al., 2016). While these particular experiences relate more specifically to the camera based approaches used in motion tracking (in contrast with body worn sensors) the discussion helps articulate a more general set of issues that relate to how we consider the design of these systems in clinical context. Indeed the purpose here is to offer an alternative perspective on these technologies that moves on from straightforward efforts to demonstrate their symptom-specific tracking capabilities. In this sense the reader is encouraged to reflect further about how other technological alternatives (sensors, tacking algorithms, information visualizations, etc.) might play out in relation to the issues raised.

2.2   TRACKING DISEASE PROGRESSION IN MULTIPLE SCLEROSIS ASSESSMENT

MS is a chronic inflammatory disease of the central nervous system. It causes a variety of symptoms, including numbness, reduction in motor strength, paralysis, ataxia and tremor (cerebellar dysfunction), as well as cognitive decline. The disease course is most frequently characterized by relapses in which the affected person experiences neurological symptoms followed by extended periods of remission in which symptoms may improve. Over time the disease can enter into a progressive phase in which a steady deterioration occurs affecting the ability to do everyday tasks, such walking or eating (Kamm et al., 2014). About 15% of MS patients have ongoing deterioration from disease onset (Hamill and Knutzen, 2003)

A large number of therapies have, and continue to be, developed to relieve the symptoms that challenge daily living and ultimately stop the progression of the disease. Tracking this progression is essential in understanding the efficacy of a specific intervention, in particular given the varied presentation of symptoms and the unpredictable nature of progression in any one MS patient. Some patients may become severely impaired over a matter of a few years, while others may live an entire lifetime affected only minimally. The measurement of motor ability is thought to be the best non-invasive indicator of disease progression. The condition is currently assessed with a standardized rating scale based on clinical examination, the Expanded Disability Status Scale (EDSS) (Kurtzke et al., 1983). Patients are asked to perform a range of functional exercises that may include movements such stretching out one arm to the side and then touching the nose (Finger-Nose test) or walking on an imaginary tight rope. The neurologist who is assessing the patient with the test makes an expert judgment about the extent of motor ability for each exercise and rates it on an ordinal rating scale (e.g., 0–4). These exercise scores are combined into a global score of EDSS that ranges from 0–10.

The EDSS is a widely used and accepted global outcome measure of MS. Of significance to our concerns it that it is the only measure approved for clinical trials by the U.S. Food and Drug Administration (FDA). As we will see later, the status of this as a clinically accepted tool has a bearing on the subsequent ways that it is acceptable to approach the design of body-tracking-based systems with a view to fitting within accepted and understood assessment practice. In contrast to newer timed measures, such as the Nine Hole Peg Test (Mathiowetz et al., 1985), the EDSS is a multidimensional measure spanning all seven functional systems of the brain. However, it suffers from low intra- and inter-rater reliability making disease tracking unreliable (Hobart et al., 2007). Further, it takes approximately 1 h to complete, and the expertise needed to carry out the examination requires it to be done by a specially trained neurologist. For this reason, it is an expensive test that is usually done yearly. Here we can see how cost factors of particular assessment procedures can realistically impact on their frequency of deployment across the trajectory of a disease.

2.2.1   CONTEXTS AND PRACTICES IN MS ASSESSMENT WITH THE EDSS

In thinking about the development of a more reliable and low-cost body-tracking system for MS assessment, it is first worth reflecting on some of the practices found in current EDSS assessment practices. The aim here is to articulate notable features of these clinical scenarios and the settings in which they occur that may impact on the ways a body-tracking-based solution can be designed to fit within these practices. While understanding existing practice is a feature of any user-centered and participatory design approach, one of the aims in the example here is to highlight the ways that seemingly unrelated features of the environment and scenario come to bear intimately on many specifics of the body-tracking system.

In a typical assessment, then, the session begins with the neurologist asking the patient about their current state. Based on the details of this discussion, the neurologist will then guide the patient through a number of simple exercises. While the EDSS assessment contains tests for all possible affected areas of the brain (e.g., eyesight), the majority of the assessment time focuses on exercises for the assessment of motor and sensory ability. Notably, the chosen exercises are varied and may involve the patient lying down on the patient bed, sitting on its edge, standing, or walking. All of these are very different types of bodily performance with features/areas of the body coming into focus through the chosen exercise. Already we can see here how different orientation of the body, different locations of the body in the assessment room or different features and shapes of bodily action might present different challenges to body-tracking systems that might be configured or positioned optimally in relation to framing and capturing certain kinds of body movement. We will elaborate further on this point later in the chapter.

What is significant in the procedural application of these exercises in different patient assessments is how the selection of specific exercises was contingent upon on a suspected pattern of symptoms for that patient. We see this issue illustrated in the contrasting requests to perform and enactments of a "standard" Finger-Nose test—by the same clinician across different patients. In the first instance, the clinician asked one patient to raise her arms to the side to shoulder height to test *truncal ataxia* and from this position to touch her nose to assess *intention tremor*. The same clinician asked another patient to raise his arms in front shoulder height to perform the *Romberg* test to assess strength and from this position asked the patient to touch his nose to test *intention tremor*. Of significance here is that the Finger-Nose test is used in both cases, but performed differently in term of the trajectory of action. In the first instance the movement began with the arm stretched out to the side and in the second instance to the front. From the perspective of the clinician in terms of what they are looking to ascertain from the Finger-Nose test, these instances are, for all intents and purposes, identical. Yet, if we consider these movements from the perspective of how they may be sensed and represented by a particular body-tracking system, the different trajectories of these clinically similar performances might appear quite distinct. In related concerns, we also saw how movements, rather than being defined by a rigid adherence to a prescribed pattern, were modified to accommodate a patient's abilities. For example, a patient whose lower body strength had been affected by the disease was not able to bicycle his legs, so he was asked to just lift the leg instead to demonstrate level of strength. In another instance, the clinician found that the patient could not lift his leg either, and she quickly adapted the exercise, asking the patient to push his leg against her hand. This variation enables a clinician to construct an individualized assessment, but is highly problematic for a computer-vision system, as large amounts of data are needed to train the algorithm for each separate movement.

This tolerance to variation raises an important consideration then. In part, this is an issue of classification and the differing bases on which one might choose to draw relationships between "*standard*" movements—and indeed the different abilities to sense the bases of these related patterns. But it will also have to do with the ways in which a clinician might be able to understand how particular components and properties of a movement might be ignorable or not in any motion classification work performed by a particular body-tracking machine.

Moreover, in constructing the assessment, the clinician is in continued dialogue with the patient; he is combining elicited knowledge from the patient with prior knowledge of the patient and additional observations. Being able to do this on the fly, the clinician demonstrates an understanding of how forms of movement may be revealing of symptoms. These abilities of a clinician also assumes knowledge of how such changes will have a bearing on the assessment scores. In particular, the clinician demonstrates how this bearing embodies the clinician's assumptions for how a score may impact subsequent courses of clinical action (Hartswood et al., 2003). Machines do not have this knowledge when assigning symptom scores.

2.2.2   CHALLENGES AND CHARACTERISTICS OF ASSESSMENT ROOM

Unlike specially designed gait assessment laboratories, the assessment rooms in which these neurological examinations were conducted had various characteristics that presented challenges to the design of body-tracking assessment tools. Of particular significance here is the small dimensions of these rooms as well as the presences of large amounts of clinical paraphernalia and clutter. Most of the rooms contained a patient bench for assessment, a sink for hand hygiene, a desk with computer for the health professional, and a cabinet for test equipment and artifacts (Morrison et al., 2016). It is also not uncommon for some patients to come into the assessment room in a wheel chair.

The small dimensions of the room combined with the clinical clutter makes these assessment rooms spatially very squeezed. As a result, the patient and any potential sensing system may need to be continually adjusted between movements. Movement performances have to be conducted around the clutter with the spatial demands of particular exercises requiring that they sometimes need to be performed in different locations around the room to avoid obstacles. For body-worn sensing technologies such *compensatory positioning* around the clinical setting might not be so problematic. With camera- and vision-based systems though, such compensatory positioning around the clinical environment would requires continual renegotiation of the relationship between sensor and body in order to achieve appropriate *line-of-sight* and *framing*. If we consider full body walking tests that make up the assessment work, these typically cannot be performed in the constrained space of the assessment rooms. For these aspects then, patients will typically be asked to move into a more expansive area of the hospital or clinical center—such as busy hallway—where the movement can be more effectively performed. While this may seem a relatively trivial matter, it places interesting demands of mobility on the sensing systems—in particular if they are camera based rather than more flexible *untethered* body worn sensors.

These characteristics of the environment and the spatial demands of the actions actually raise some interesting challenges for various body-tracking systems. For example, the clutter and paraphernalia within the setting, are not just a challenge for achievement of appropriate line of sight between moving body and sensor or appropriate framing of the moving body by the sensor, but also create additional challenges in the ability for a system to successfully parse the body from the rest of the foreground and background environment. The visual element of clutter has the potential to interfere with raw sensing signals from a depth sensor. A metal chair behind the patient, for example, cannot be sensed by a depth-sensor and will leave a large black spot in the depth video. The sanitation paper for example, if hanging off the bed and behind the patient's legs, will make it difficult to segment the patient from the background. Such constraints may be further exacerbated by the particular proxemic demands imposed by the minimum distance requirements of various body-tracking systems. The depth sensor in the Kinect, for example, requires the patient to be between 3 and 6 m away to be consistently tracked. Such proxemic demands of a body-tracking system present challenges to the ways by which the sensed and sensor can be configured within the confines of typical clinical settings.

Additional challenges arise in these settings because the spatial constraints and artifact obstacles introduce a further variability into any potential data captured. Patients may need to perform *compensatory movements* with respect to any spatial constraints and obstacles. For example, in observations of clinical sessions we note that patients would take twists and turns to avoid obstacles in busy hallways while carrying out walking tests in the clinical environment. The challenge here is that while such compensatory movements are an important practical feature of performing exercises within these environments, they introduce unwanted variation into the performance. This variation is not something that is a problem for a clinician in assessing symptoms, but can result in highly variable sensor data. This then poses certain challenges for machine learning-based analysis of disease-related movement comparisons.

Finally, these assessment spaces are part of a larger clinical system and workflow that extend beyond the immediate requirements of a specific assessment session. In many of the sessions that we observed, for example, interruptions of various sorts were commonplace meaning that there was rarely a continuous flow in the clinical proceedings. Other clinicians would pop in and out of the assessment room to ask a question or to borrow some piece of equipment. Likewise, the telephone would frequently ring. Any standardized mechanisms for capturing as well as classifying trajectories of movement behavior would need to be robust to the idiosyncrasies of an interruption-based environment.

2.2.3   DOCTOR-PATIENT RELATIONSHIP IN ASSESSMENT

A key area of concern for us in understanding the role of body-tracking systems in healthcare is its consideration as a collaborative enterprise. In this respect, the moving body of the patient cannot be regarded as an isolated entity to be looked on by the doctor or to be sensed and assessed by a body system. Rather, it is important to consider the moving body of the patient in the context of a collaborative doctor-patient relationship. There are a number of important and interesting concerns here that may pertain both to the practicalities of potential sensing and parsing by a computer system but also the ways in which elements of the system interaction may fit within the broader considerations of the doctor patient relationship.

Let us consider, for example, some of the practical manifestations of neurological conditions such as MS. Patients at particular stages of disease progression will experience difficulties with balance and strength and as such have the potential to fall as a consequence of their motor difficulties. When performing standing or walking exercises in the context of EDSS assessment, our observations revealed how this was difficult for some patients to do unaided. In these circumstances, a clinician or a family caregiver accompanying the patient will step in to physically hold and support the patient. Other aspects of the consultation, too, brought further contact between clinician and patient. For example, it is common for a clinician to touch the patient so as to indicate which side to attend to or to assess other physical attributes of movements such as their kinetic rather than kinematic components.

On the face of it this seems like an entirely practical and reasonable thing for the clinicians to do and in the context of current assessment procedures has no bearing on the observational judgments of movement performance. Yet these moments of contact have significant implications for the use of body-tracking technologies as clinical assessment tools, in particular those approaches that utilize camera- or computer-vision. The reason for this is that in offering support during walking or standing, or when indicating parts of the body to be used in various exercises, the clinician and patient body come together making their visual delineation by tracking technologies a challenge. In particular, this is the case with camera-based tracking and computer-vision approaches which attempt to algorithmically delineate the moving body of the patient from other people and environment artifacts.

In addressing these concerns one might consider the deployment of sensors for which such confusion may be less of a concern. In part, one might also look to address these with improved computer-vision techniques that more explicitly attempt to solve the problem of *patient body delineation* in these contexts. In lieu of meeting these particular computer-vision challenges, although, an alternative approach is to acknowledge that such collaborative coming together of patient and clinician is a practical reality. Such acknowledgement may then involve ways to facilitate the understanding and awareness of interacting bodies by the patient and clinician and the impact these interactions may have on any computer-vision assessments of motion performance.

Other practical features of the clinician patient-interaction include things such as the instruction and demonstration of *correct* exercise performance. Again, doing this is a collaborative and embodied concern whereby the clinician needs to physically configure themselves with respect to the patient and other artifacts in the room so that the demonstration is visible and understandable. The achievement of correct bodily performance by the patient is something that is collaboratively negotiated and realized by the doctor and patient together through a combination of verbal, visual and tactile instruction.

Critical here is that the demands of these clinician-patient configurations have to be combined with other orientation demands of clinical artifacts. As highlighted in the work of patient-clinician interaction by CSCW researchers, e.g., (Heath, 2003; Greatbatch et al., 1993), the ability of the patient and clinician to choreograph their bodies and attention with respect to each other is profoundly affected by the material form of other documents and artifacts that may be present in the interaction. Introducing a new body-tracking system with accompanying visual displays into this doctor-patient interaction is something that needs to be considered in terms of its impact on the *in situ* organization of these encounters. In the case of demonstrating correct exercise movements, how does a clinician stand in front of the patient to demonstrate movement without blocking the line of sight between a sensor and the patient being tracked?; how might a clinician conduct these demonstrations if required to interact with a screen on the assessment system?

Extending our viewpoint from thinking about the patient body as an isolated concern to be tracked to one that is an active part of a larger social interaction reveals further features of note within the consultations. An interesting feature here is how assessment and its outputs are not just a concern for the clinician but also for the patient. What we mean here is that such measures of movement capabilities become a resource for the patient to understand their disease progression. The measures become objects that can be shared with patients in discussing improvements or decline in motor ability. Indeed these can become subject to negotiation with the patients as to whether they represent an adequate reflection of the status of the patient's everyday experiences with their motor ability. That is, there is sometimes additional negotiation with the patient or caregiver as to what a particular level of performance might be. Mentis et al. (2015, 2016) referred to these intriguing facets of the clinician-patient interaction as acts of *co-interpretation* in which patients, clinicians, and carers discuss and contest data on the way toward a collaboratively achieved assessment.

What this leads us to question is the status of movement data in the clinical process of assessment. Rather than just an objective quantification of movement performance being the end goal here, the aim might rather be framed as one of supporting different acts of interpretation either by the clinicians concerned or the patients in collaboration with the clinicians. This framing then opens up other aspects of the design challenges associated with these technologies, namely how might the movement data be *represented*, *visualized*, and *manipulated* in support of these interpretive acts. The work of Heath and colleagues (e.g., Heath, 2003; Greatbatch et al., 1993) in the area of clinician-patient interactions is informative here. Of significance in their studies is the embodied nature of these interactions and in particular how these encounters become configured with respect to the material properties of clinical documents and artifacts. While some of their work was concerned with the move from paper to digital documents in these encounter, the work still highlights broader need to consider the material (both hardware and software) representation of captured movement data in the interpretive acts of assessment.

2.2.4   SUMMARY

Before we move on to consider some aspects of system design rationale let us briefly summarize some of the key points that we might want to consider in a contextual understanding of the clinical assessment context. First, clinicians adaptively construct assessment workflow in accordance with the status of patient symptoms and movement abilities emerging and elicited during the interaction with a patient. In this respect, clinicians accommodate high variance in the performance of movement exercises in the construction of their assessment workflow. So while movement tests may be *standardized*, what that implies for a clinicians in eliciting symptomatic clues may be different from what it means for a potential computer-based system designed to classify particular patterns of movement. In light of this, in developing our own system, it was felt that a *passive* capture system that "watched" during a normal assessment would be unlikely to work successfully because of the enormous range in movement performance. Rather, in was felt that such circumstances would best be served by a system that was based on an active, deliberate performance of specific, standardized movements.

Related to this is the idea of different notions of assessment and where best to situate body-tracking technologies. Supporting the *in situ* organization of a clinician's assessment of disease progression may be different from supporting an objective demonstration of disease progression in a clinical trial. Part of the skill of the clinician is in the adaptive administration of movement tests in relation to patient symptoms rather than being overly prescribed in their administration. The former then may be less tolerant to a rigid and inflexible specification of movement protocol. By contrast, in the context of clinical trials, there may be a need for a more standardized administration of movement assessment protocol. While adaptation may still be necessary, the skill of the clinician in directing movements is not utilized. Other manual tests, such as the Nine-Hole Peg Test (Mathiowetz et al., 1985), are administered by nurses. These alternative contexts provide inspiration for where else a sensor-based system may sit within the clinical context.

Our observations have also shown that the assessment is influenced and configured in relation the physical space in which it is carried out. Clutter in the background of the depth videos is something that must be practically managed. The mobile nature of the assessment, due in part to making space for movement performance around clutter, must be accounted for with appropriate positioning while recording. *A view onto what the camera "sees" is needed to help the health professional manage the environment.*

The assessments are a collaborative concern in which the patient and clinician work together through verbal and physical instructions to produce an assessment. This is in talk, but also in the ways the bodies of both patient and clinician work together as a combined entity in the production of the patient's body as a meaningful object of assessment where clinicians move dynamically in relation to the patient, often touching and supporting them. Systems need to consider features of the clinician-patient interaction, both as a practical concern in the organization of successful measurement but also in the ways the system may facilitate acts of interpretation and co-interpretation.

2.3   UNDERSTANDING CONCERNS IN SYSTEM DESIGN: ASSESS MS SYSTEM

Let us consider how these above issues play out in relation to the design of a specific system. In this particular instance, we draw on the work of a particular system called Assess MS, a depth sensor-based body-tracking system designed to assess and classify motor ability in MS patients (Morrison et al., 2016). To illustrate how we addressed these design issues, we will first describe the final system as a reference point, and go on to discuss various considerations at play in the design decisions and alternatives. This discussion will help highlight the complex interplay between particular features of the design and key aspects of clinical workflow and context in movement assessment.

2.3.1   SYSTEM OVERVIEW

The Assess MS system uses depth-sensing computer vision and novel machine learning algorithms to determine the appropriate symptom score on the EDSS for specific motor symptoms (e.g., ataxia). One of the key reasons for using depth-sensing cameras as the basis for body tracking in this instance was to offer the potential for unencumbered, full-body tracking without requiring particular modifications or additions to the patient (e.g., adding markers, wearable sensors, or changing clothes). The aim here was to require minimal set-up and training such that a system could be practically deployed within a traditional clinical environment as opposed to a bespoke body-tracking laboratory. Part of the aim here was to make it usable by nurses, as well as neurologists who are the traditional assessors. Assess MS has three interconnected components: a system for capturing depth videos in the clinical setting, a set of bespoke algorithms for pre-processing the depth-videos and classifying them; and a set of movements that highlights clinical symptoms. An interface that visualizes the results will be added at a later date. We describe these components in detail below.

The system is shown in Figure 2.1 has a 21" patient-facing screen used to instruct patients in the assessment movements. A smaller tablet computer with touchscreen capability is mounted on a mobile arm joined to the back of the unit. This interface is used by the health professional to position the patient, select the assessment movements to be performed, and complete the recordings. A remote control enables the health professional to move freely around the room to support the patient as needed. The whole system is mounted on wheeled legs for ease of maneuvering with the Kinect mounted on top.

The *patient interface* provides a standard presentation of assessment movements through animated instructional videos. The animations consist of simple line drawings accompanied by verbal descriptions. A red border and beep indicate the start of a recording and a green border and chime, its finish. The name of each assessment movement is also displayed.

The *health professional interface* provides a number of navigational devices. The health professional can play a movement instructional video, or begin a test. Each page contains one button, which enables the beginning of a test, recording of a movement, or stopping of a recording. Arrows at the top of the interface enable the health professional to skip movements (e.g., Finger-Nose test) or variations of movements (e.g., left side, eyes open). Movements can be repeated by skipping backward. A navigational bar at the bottom shows visually which movements have been captured and which skipped. Protocol instructions, such as perform with feet on the floor, are written below the videos.

Two tools were provided for managing the camera view. A positioning window provides a view of the depth image stream with center crossbar to which the patient should be aligned. It is available before the sitting and standing components in the test as a full-screen feature and in a persistent window in the upper-right hand corner throughout. The distance of the person from the camera is indicated below the image. We also provided a screen to help health professionals identify clutter. This screen visualized areas in the depth image that consistently reported zero values, turning them green. This might be a chair left behind the patient, or a handbag left next to the patient on the bed. This screen is shown to the health professional before the assessment begins.

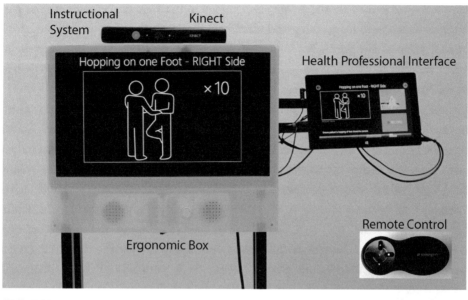

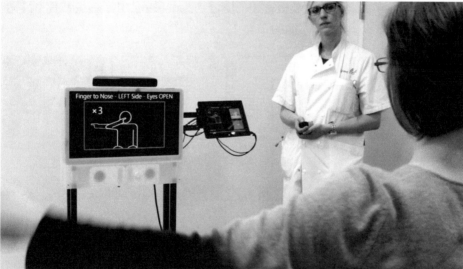

Figure 2.1: The Assess MS System—(top) component parts; (bottom) with clinician and patient.

2.3.2 ALGORITHMS

Often when considering body-tracking technologies in healthcare contexts much of the discussion emphasizes characteristics of the sensing technology itself and how these may pertain to body-tracking capabilities as well as alluding to high-level practical concerns about their deployment. However, over and above the sensing technology in itself are concerned with the particular algorithmic approach used for tracking and classifying movements. What we will see later is how algorithmic decisions have significant impact on the organization of action in interactions. With this in mind we want to highlight aspects of the algorithms used in the Assess MS system so as to illustrate the implication of these choices in later discussion.

In this particular instance a supervised machine learning approach has been taken using clinician-provided EDSS symptom (sub-) score labels. Customized randomized Forests and novel ensembles of randomized Support Vector Machines are used to discriminate landmarks in the depth videos that contribute to the classification of whether person is a healthy volunteer or a patient. Depth-videos are pre-processed to provide foreground segmentation and registration; nearest neighbor in-painting is used to fill missing data and template-matching in depth space is used to detect and center the head. Movement is captured through the calculation of optical flow throughout the video. Changes in direction in optical flow are used to threshold the data and isolate oscillating movements. Features are then stochastically sampled from spatio-temporal cubes in the depth video and compared. No temporal alignment of the videos is carried out as temporal information may carry important information about disease state. The sensor data pipeline for pre-processing the data is shown in Figure 2.2.

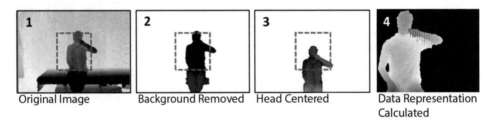

Figure 2.2: Sensor data pipeline for Assess MS project.

Within the Kinect SDK, an algorithm is provided that automatically infers where the major body parts are and connects them. This is represented in the form of a "skeleton" that is overlaid on the image of the tracked body. Significantly, this algorithm, which is designed for the specific context of gaming and system control, is not used in the Assess MS system. At first glance this might seem like a curious decision but while this "skeleton" provides an off-the-shelf way of doing the pre-processing of the depth videos and capturing movement, the inferences made about *relevant* movements are not well suited to this clinical context. A key issue with the "skeleton" here lies in the use of spatio-temporal filters to the movement data with a view to removing *jitter*. The use of these filters in commercial off-the-shelf systems helps *compensate* for inherent limitations in the depth sensing aiming to smooth out the inferred movement in ways that renders the motion tracking in suitable form for use as an interaction mechanism. However, some of this jitter still remains. One of the challenges here lies in the ways that particular algorithms may deal with jitter: the extent to which it may conceal pathological tremors or the ways that it renders it problematic to distinguish between pathological tremor and jitter. Furthermore we found that the skeleton could not be post-rendered consistently on the depth videos of movements captured in the clinical environment. Specifically, a patient sitting on a bench with a wall close behind made it difficult to initialize the algorithms. Another issue with the skeleton concerned its ability to deal with certain forms of movement exercise that are designed to elicit particular symptoms. For example, with the inferred "skeleton" provided by the Kinect, the physical touch of two body parts (e.g., in the Finger-Nose test) would cause an error just at the nose when the pathological aspects of the movement are strongest. Here we can see an interesting example of where movement design for clinical concerns of motion assessment may conflict with the performance attributes of particular algorithmic approaches in body tracking.

2.3.3   MOVEMENT EXERCISE PROTOCOL

The movement protocol for the systems contained 11 movements, which are depicted in Figure 2.3. Six movements were chosen from the EDSS assessment to cover the function of the upper and lower extremities and the trunk. Two activities of daily living (ADL) movements, Drinking from Cup and Turning Pages in Book, were also included. In addition, three new movements were defined, Finger-Finger test, Drawing Squares, and Rotating on the Spot. These were created to capture in a potentially more camera-friendly way upper and lower extremity function. Unlike other more recently developed tests, such as the Multiple Sclerosis Functional Composite (MSFC), the aim was to maintain the specific information on motor ability provided by the EDSS symptom (sub) scores.

In thinking about the movement design for sensor based assessment a number of additional considerations are at play. One such issue relates to the clinical legitimacy of the movements that are assessed. The standardized movement protocols of the EDSS for example, have established themselves as a widespread clinically accepted standard. Part of the consideration of any new system is how it fits within such a validated, established and understood process for classifying the clinical symptoms of MS. In spite of the articulated shortcomings of these standardized movement protocols, there is considerable value in the ways that they are embedded in clinical practice. With this in mind it became imperative in the design of Assess MS to work with movements that were aligned with the established movement protocol of the EDSS. The movement protocol, then, was strongly guided by the conceptualisation of the tool as a replacement for part of the EDSS. Active capture is done for all movements, with the majority being drawn directly from the EDSS. This would ensure that any inferred classification by the system would have significance within established practices of motor assessment.

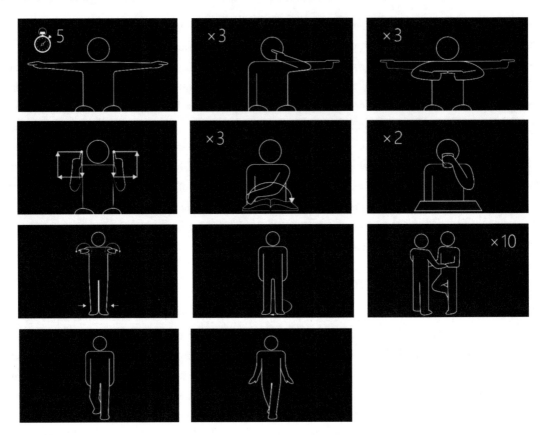

Figure 2.3: Neuro-assessment movements performed with Assess MS: (top-left to bottom-right) Ataxia, Finger-Nose, Finger-Finger, Drawing Squares, Turning Pages, Drinking from a Cup, Romberg, Turning on the Spot, Hopping on One Foot, Normal Walking, Tightrope Walking.

Such a choice of movement grounded in traditional clinical practice, while important, also poses significant challenges for body-tracking systems since the movements are designed with different criteria in mind. EDSS movements were developed to highlight symptoms to a neurologist not a camera, making capture difficult. For example, the Finger-Nose test was difficult to standardize, as clinicians frequently did it in different ways. With a view to offering more machine readable movements, designers of the Assess MS system had originally proposed a more game-like movement interaction that could guide movements in specific ways. One example of such a proposed movement would be a simple game in which a patient had to pop each balloon as it appeared. From a camera and computer-vision point of view such a movement would facilitate a precise delineation of the start and end of the movement as well a precise expected trajectory. However, this kind of movement was deemed unacceptable owing to the challenges of convincing the medical audience that the movements elicited the symptoms of interest in the same way as the EDSS. This would not be questionable if movements from the EDSS were used. Further practical issues need to be taken into account when thinking about depth sensing "friendly" movements from the EDSS set. For example, the lying down movements were excluded in the Assess MS system. Similarly, side-view leg lifting exercises were excluded, as the wearing of skirts or baggy trousers would make it difficult to judge, as the leg could not be isolated.

2.3.4   ENSURING STANDARDIZED MOVEMENT PERFORMANCE

One of the key challenges to address in the design of these systems is how to ensure that the movements captured are standardized so that any variance in movement captured can be assumed indicative of disease state. Designing to support the instruction of "*correct*" performance of a movement is not a simple task. Some inspiration can be drawn here from the literature on virtual reality, specifically studies of design strategies to teach people movement using depth-sensing computer vision. Many systems place a person in a virtual learning environment with a teacher. In some studies the student's movement is presented alongside or overlaid on top of that of the teacher's with either a skeleton or mesh representation (Bailenson et al., 2008; Chan et al., 2011; Chua et al., 2003, Usui et al., 2006). An alternative approach has been to prompt which direction the arms or feet should go, leaving the path of movement to be decided by the user (Charbonneau et al., 2009).

In speaking with clinicians however, movement instruction in the case of clinical assessment may be more constrained than in other domains. Initial approaches explored in Assess MS included showing: (1) the patient's depth video next to the instructional video; (2) annotations on the depth video to highlight important aspects (e.g., arm height); and (3) movement traces of the patient on the instructional video. While such approaches seem reasonable from a purely instructional point of view they are deemed unacceptable by clinicians. The argument here is that showing patients a visual stimulus and feedback designed to correct movement *during* performance would potentially impact of the use of the movement in neurological assessment. With this constraint in mind, it was only possible then to provide a presentation of the movement before it was performed, substantially constraining our options for movement performance standardization.

This led to an eventual instructional solution that presented each movement with a simple animation to help ameliorate the idiosyncratic instructions of health professionals or guide those not skilled in performing the EDSS assessment. An important consideration in the design of these animations was their position is the clinician-patient relationship and the flow of the consultation. Animations were provided to support both the patient and the doctor in their guidance of the patient. Animations were presented on both the patient interface and the clinician interface. Initial versions of the animations included prompts for protocol instructions, such as lowering the bed, in the instructional videos. While it did server the purpose of aiding the health professional to remember protocol instructions, it also had the unwanted consequence of doubling the length of the assessment. It is interesting to note here that by focusing on a particular issue of design in isolation, it is easy to lose sight of the broader consequences. Doubling the length of the assessment sessions is not simply frustrating for both clinician and patient but also has financial implications that ultimately result in greater inter-assessments intervals—one of the very issues that the low-cost ubiquitous sensing paradigm is trying to avoid in the first place. With this in mind the instruction animations were reduced to a maximum length of 15 s and shorter where possible. Having the ability to skip instructional videos also brought with it a greater sense of control over the procedural use of the system by clinicians allowing them to adapt to the contingencies of clinical work. There is also an assumption here that after they do it the first few times they will be better able to quickly do it again. So, that is an upfront cost in terms of time but not an ongoing cost.

Positioning the clinician in relation to the patient is also a key element of instructional work in the clinician-patient interactions. As we have discussed earlier, one of the key challenges for camera-based body-tracking technologies is the potential for line-of-sight interference from other bodies in the assessment context—in particular the clinician. The Assess MS orients to this concern in an interesting way through design of the physical device and, more specifically, the relative orientations of patient and clinician screens with respect to each other and the sensor. With the sensor and the patient screen oriented toward the patient, the clinician screen was oriented to the side or back. In this respect, through the need to interact with the screen, and because they were sensitive to not blocking the patient's view of the instructional screen, the clinicians were naturally positioned out of the line-of-sight of the depth-sensing camera. A remote was provided on the occasions that they needed to interact with the patients.

While these aspects of the system design orient to the concern of standardizing movements for body-tracking-based classification across patients and across time, there remain fundamental questions as to what constitutes acceptable standardization. As we saw earlier in the discussion of existing assessment practices, the performance of a nominally same exercise is carried out in a wide variety of ways in terms of the spatio-temporal trajectories of particular actions. While clinicians are tolerant to this variance, and even contingently encourage such variance in the performance of an exercise, it is not clear whether the basis and criteria for movement classification by clinicians are the same as required by a body-tracking system. An additional concern here is the extent to which a clinician without sufficient knowledge of the underlying algorithms can understand the basis for tolerance in the body-tracking-based classification of related movements. That is, how does a clinician achieve system friendly standardization of movement exercise performance? We go on to discuss this further in the next section.

2.3.5   FRAMING AND STANDARDIZATION—SEEING HOW THE MACHINE SEES

Over and above movement standardization in the context of clinical assessment, there is also a practical concern with standardization of any captured representations of these movements. Standardizing the representations of particular forms of motion is the basis for any analytic comparisons and classifications to be performed on the data. As a simple illustration here, imagine a movement exercise that is performed in a completely standardized way. If we capture the silhouetted body from the front when performing this exercise, it would look very different from the silhouetted body captured from the side. That is, while the performance of movement in this scenario is identical, the captured silhouettes would look dramatically different. What is being illustrated in this example is how any capturing of consistent and standardized representations of movement patterns presents a practical problem for the clinician and patient in everyday movement assessment contexts. As we will discuss, the situation is more complex than the illustrative example suggests, but it helps draw attention to this facet of the problem that needs to be supported in the design of the system.

Part of the challenge arises in the need to capture a diverse set of movements with a single sensor set-up, yet there are different demands for different forms of motion. If we consider walking exercises in gait analysis, for example, the relevant features of the body in motion are best captured by mounting a camera lower (e.g., at knee height) or higher (e.g., above a door); for upper-body movements capture would be best done with the camera around 1.6 m from the ground. The physical environment also caused difficulties that relate in interesting ways to characteristics of the sensor. Gait, for example, could not be captured in most of the clinic rooms because of their constrained size. The Kinect camera requires people to be 3 to 6 m away from it in order to get a reasonable image. In most clinic rooms, even big ones, the furniture did not allow such distances between the camera and the patient to be practically achieved. As a result, gait was frequently captured in the hallway. This seemingly sensible solution from the neurologists' perspectives introduced issues of highly reflective spaces as in Figure 2.4-right along with the clutter of other people and things.

A key aspect of managing these framing challenges was support for the mobility and maneuverability of the device. This allowed the device to be positioned appropriately with the necessary flexibility to accommodate the range of exercises as well as the clinical paraphernalia and people comprising the assessment environment. Maneuverability of the device though is only part of the story. Additional support for this framing was provided on screen with a live feed of the depth image and a center crossbar to which the patient should be aligned (Figure 2.4-left). Importantly, positioning is something that needs to be done not only between sets of movements (e.g., sitting and standing), but throughout the entire examination due to the large amount of movement. In the Assess MS system, the positioning window was made *persistent* in the upper-right hand corner of the interface to provide ongoing feedback through the examination. These support tools are important for real-world use of these systems because they help health professionals make standardized data possible in non-standardized environments. They allow the clinicians to: adapt to different clinics, e.g., move the camera to a new location for movements which required props; manage unexpected happenings, e.g., a person walking by or a hand-bag on the bed; and manage the interaction with the patient, e.g., adjust the camera if the patient had shifted on the bed. These tools enabled the health professionals to adapt to the unexpected in real-world use.

Figure 2.4: Data variance from the environment. (left) Limbs out of the camera view; (center) clutter behind the patient; (right) reflective spaces

As we have raised a number of times, the conduct of the assessment is not something attributable to the clinician alone. Indeed the correct positioning of the patient is a collaborative effort. The patient too does significant work to achieve correct positioning and performance. An important consideration in the design then is the provision of ongoing positioning feedback to the patient. That is the patient needs visible access to this information also. In the case of Assess MS the system utilized the dedicated patient screen for these purposes so that it could be oriented to the patient independently of the clinician who needed to stand out of frame.

In the context of sensor-based body-tracking technologies, the standardized framing challenges are actually a more complex concern and extend beyond what one might consider a straightforward camera view-finder model. The key to thinking about the design of these technologies is the notion that computer-visions systems do not "see" as people do. Much of what is important in the standardized framing challenge is an orientation to the ways that the system "sees" and not just what the clinician sees. While computer-vision algorithms can segment and parse scenes with some considerable success, the approaches used mean that they rely on a different set of dependencies relative to the human visual system. What is obvious background to the human visual systems might be regarded as clutter to the algorithms of a computer-vision system. What is "specular reflection" to a person is a lack of data to a computer-vision system. As we saw in our observations of real-world clinical assessment environments, they are full of things such as furniture, machines, and other people. So, for example, the reflective metal chair behind the patient in Figure 2.4-center, makes the background removal in the image pre-processing stage problematic. Objects on the bed, whether a handbag put down by the patient next to themselves, or a bed sheet hanging behind the knees, can also make the distinguishing between the background and the person highly problematic. This kind of clutter leads to data that is thrown out. It therefore requires management by health professionals for the system to be workable in practice.

With these concerns in mind, an important element of the system design is the provision of feedback to the health professional and the patient that allow them to orient to the ways that the body-tracking system "sees". In the Assess MS system, for example, the feedback provided is not just about whether the patient is spatially in the frame of the camera but also whether there are features of the environment that may be leading to distortions in the captured representation of the body. These tools provide the basis for clinicians to understand how the system "sees" and as a result, manage the environment to support the computer vision. Importantly, these feedback representations need to help the clinician understand the relationship between what they are seeing and what the system is seeing.

These issues of what the computer is seeing and the "inferences" it makes are further complicated by the specifics of the algorithms used and the spatio temporal features of movements that form the basis for comparison and classification of represented movements across patients and across time. Again, drawing on the Assess MS development work, let us consider some interesting examples of this. One of the decisions made with the system was the need to split the repetitions of assessment movements into distinct entities (e.g., if the Finger-Nose test is done three times then treat these as three instances) to increase the amount of data available. In making such a decision, this has the consequence of requiring the starting and stopping points of movements to be done in a consistent manner to increase alignment across the captured video instances. In terms of what this meant for the organization of action in the clinical setting, it subsequently mattered if a patient started with their hand out or in their lap whereas previously it had not been a concern.

Similarly, requirements for appropriate positioning in image standardization were also contingent upon certain properties of the underlying algorithms. In the first example explored, data was represented as a slit scan image of the depth video, in which the video was translated into an image by projecting the closest depth in each row of a depth image, into a single column. A column would be generated for each depth image in the video (shown in Figure 2.5-left). In this image, it is essential that limbs be in view at all times to avoid black spots in the image. As such, consistent positioning of the patient in motion is key for this approach. In contrast, the second data representation explored (see Figure 2.5-right), is the calculation of optical flow in randomized spatio-temporal cubes of the video. This representation is much less affected by poor positioning, as movement is being sampled randomly and for the most part, not at the extremes of the extremities.

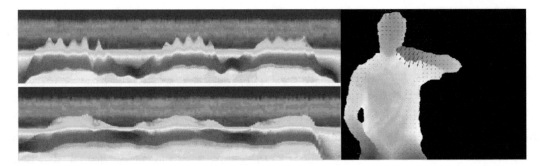

Figure 2.5: Different data representation used in the machine learning. (left) Slit-scan image of the finger nose test; (right) optical flow calculated in stochastic spatio-temporal cubes. These different data representations influenced how image capture standardization could be achieved.

The interdependencies between the technical components of body tracking and features of practical action in the context of a clinical setting are clear to see in these examples. Providing feedback that make these things explicit to the clinician is one essential component in the design of the systems but is unlikely to be sufficient. If we refer back to earlier discussions about the contingent nature of movement assessment procedures, this highlights how clinicians flexibly adapt the procedures to accommodate features of the clinical context or the symptomatic requirements of the patient. Such adaptations are done in the knowledge that any performance differences are not significant in terms of assessment outcomes. In light of our discussion around image standardization in body-tracking systems, such adaptation becomes more of a challenge since it is further dependent on having a working knowledge about algorithms and their sensitivities to particular characteristics of movement trajectories.

2.3.6   REPRESENTING THE MOVEMENT MEASURE AND CLASSIFICATION

As we have seen thus far, body tracking in the context of clinical assessment is not as straightforward as it might first appear. In developing these systems, one is not simply automating the assessment process and classification by replacing the clinician with a machine. Rather, what we are developing is something that augments the interpretive practices involved in assessment. This transition in perspective is not always readily apparent. In the early development of Assess MS, for example, the original assumptions were for the system to generate a classification number much in the same way that a clinician currently does. As such, little consideration was given over to the issue of how the measurement and classification of movement should be represented. Yet as we come to accept that algorithms are not "seeing" in the same way that a clinician sees, the issue of measurement and classification becomes more relevant. For example, the number the algorithm provided could not answer questions such as, "Did the patient have an upper disymmetria score of 2 because of tremor or ataxia?" These symptoms are scored together because of their pathological origin, but have different treatment approaches. These classifications are contextual and part of the resources of ongoing clinical interpretation.

Shifting perspective from one of measurement to one of supporting clinical interpretation brings to the forefront the important issues of how to represent and visualize the measurement and classification outputs of body tracking such that they are support the process of clinical interpretation. There is a growing literature about communicating an algorithm's decision process to the user. Research suggests that providing why and why not solutions can be informative to users (Kulesza et al., 2011), or even essential to use (Herlocker et al., 2000). Alternatively, decisions can be shown at each decision point in an algorithm, providing a platform for people to interpret a decision (e.g., van den Elzen and van Wijk, 2011; Sharara et al., 2011; Vidulin, 2014). These two approaches fit well with research that suggests that sounder mental models of the system improve satisfaction in system use and self-efficacy of computer usage (Kulesza, et al., 2012). Let us consider these concerns more concretely with examples from the Assess MS system.

The first visualization created by our machine-learning colleagues to inspect, as well as illustrate, how their algorithm was functioning is shown in Figure 2.6. It consists of a heat map of the spatio-temporal cubes used by the algorithm to distinguish between healthy volunteers and those with MS. The spatio-temporal cubes are aggregated across patients and flatten onto a single image. As a reference point, a random frame from a single patient video was used as the background to show how a body may relate to the spatio-temporal cubes. Although it was not intended for our clinical colleagues, their reactions illustrate the challenges faced when communicating results.

The relationship between the spatio-temporal cubes and the body is not obvious. Many of the cubes are off the body, such as next to the waist. From an algorithmic point of view this makes sense: these boxes would be expected to contain movement when the patient has ataxic symptoms in comparison to no movement when the person is healthy. The interior of the body is less likely to show differences between patients and healthy volunteers. However, when a health professional assesses a patient, they look at the body and not where it might go in space. This is a substantial disconnect which made it difficult for the health professionals to link the algorithm's decision process to their own.

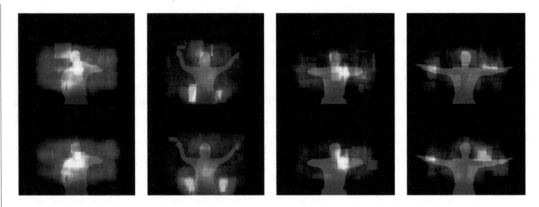

Figure 2.6: Heat map visualization of the spatio-temporal cubes that contributed the algorithmic decision process to discriminate between healthy volunteers and those with MS. Yellow are the most important decreasing in intensity to light red as the least import

A key issue was the use of spatio-temporal cubes as the data representation. One of the main motivations for choosing this representation was that it was agnostic to particular movements. This would enable a single approach to processing any movement, an advantage given the large number of movements in the Assess MS protocol. A generic approach also facilitates the extension of the algorithm to other movements or conditions. While the choice is well motivated from a machine learning point of view, it creates a challenge for clinical interpretation purposes. A single data representation cannot relate directly to any movement and thus to what the clinician may be accustomed to seeing. An unanswered challenge is how to create a shared representation that is useful to an algorithm, but relates to the body in ways that are interpretable by the clinician and patient.

Our explorations of how we might map the features used in the algorithmic decision-making onto a representation of the body raised further discrepancy with machine learning optimal choices. Figure 2.7-top shows a slit scan image on the left in which it is possible to see the three repetitions of the finger-nose test and, in the bottom image, visible tremor is. We mapped the heat map in Figure 2.6 onto the image in Figure 2.7-top in order to get Figure 2.7-bottom. In doing so, we discovered that the spatio-temporal cubes used often spanned the majority of the movement. This makes sense for a cyclical movement, as a longer cube will have three times as much information. However, this makes it difficult to match which part of the movement is relevant in the decision-making process, a key to clinical understanding.

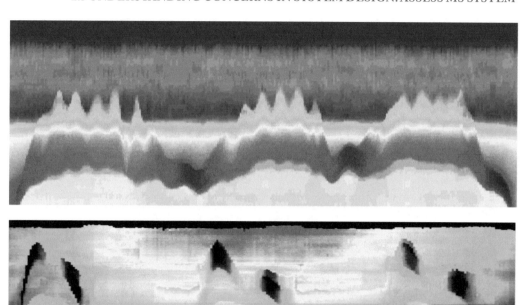

Figure 2.7: Slit scan image of finger nose test: (top) image shows tremor; (bottom) feature heat map overlaid onto top image showing no association between features and movements.

The above explorations looked for common ground between clinical understanding and machine learning by focusing on how the latter could be related to the body. An alternative approach would be to focus on what machine learning can do that clinicians cannot. In this case, the machine-learning algorithm can provide a patient cloud with some sense of which patients are most similar to the new one, as in Figure 2.8. This can enable visualization of a more granular scale as well a prognostic view of disease state that could potentially help with treatment choices.

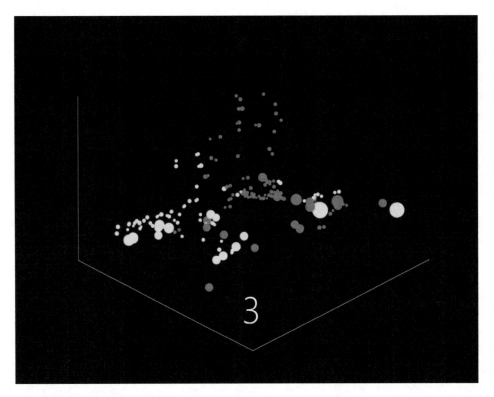

Figure 2.8: Communication of algorithmic result through a patient cloud.

2.4   CONCLUSIONS

What we have seen in this chapter is a glimpse of the significant potential for use of low-cost ubiquitous sensing technologies in the assessment of motor disability symptomatic of neurological disorders. While there has been early work exploring the feasibility of these systems for clinical movement assessment, there is still relatively little work that explores what it takes to make the transition from lab-based feasibility to clinical practice and environments. In making such a transition our attention needs to turn to a broader set of characteristics of particular ubiquitous sensing technologies beyond those that concern their demonstrable ability to track and measure bodily movement.

In attending to these concerns and illustrating them in the context of a system development, we have begun to see some of the interdependencies between system features and potential features of the clinical context. Some of these pertain to clinical symptoms themselves and how these are manifest in specific types of movement. Many assumptions about the relevance of certain movements may be bound up in body-tracking approaches with different sensors and approaches inadvertently or deliberately keeping or discarding features of movement. As well as understanding the movement assessment requirements of particular conditions and their symptoms, the physical requirements and sensitivities of these sensors come to affect the ways in which action in the clinical environment can be organized. We see this play out not just in terms of the environmental constraints on sensor performance but just as significantly in terms of how sensor and system characteristics can influence the social organization of action in these contexts.

It is striking to see how a complex chain of dependencies can link the specific symptoms of movement disorders to the details of algorithm choices in body-tracking systems for clinical assessment. For example, we saw how patients with clinical movement disorders, may at times lack stability and require physical support from a clinician or carer, which in turn brings two bodies into contact. This, in turn, can create challenges for certain camera-based sensing technologies and algorithms that may depend on the separation of bodies in space for successful tracking. Conversely, the existence of these sensor characteristics may lead to reconfigurations of behavior to accommodate these limitations. In this illustrative instance, however, we can see how such accommodation may hinder the ability of the clinician to offer appropriate support required by these patients. The point here is that bringing these body-tracking technologies into real-world clinical contexts may require accommodation on the side of technology design and on the side of work practice reconfiguration. In determining what is appropriate here we need a much stronger articulation of the mutual dependencies of clinical, social, and technical.

This articulation entails certain shifts in perspective. One important shift that we have tried to articulate here is the consideration of body tracking in assessment as a collaborative concern. This shifts our thinking from one in which there is a primacy of objective body measurement as a central concern to one that acknowledges the centrality of assessment as a collaborative practice. Of significance here is the acknowledgment on the clinician-patient interaction in this work. The challenge here is not solely with the development and refinement of complex machine learning algorithms but with the ways that these systems allow a clinician to manage an interaction. As we have seen this is in part concerned with the practical organization of patient behavior in the achievement of standardized movement performance. But there is something more at play here that relates to interpretive aspects of assessment that is collaboratively achieved among clinicians but also, significantly, a co-interpretation between patient and clinician. Orienting to these concerns is also a key area for future system design in real-world instantiations of these systems. This points to, for example, a need to consider not just sensing as an input to motion assessment systems, but also the ways in which data might be visualized and represented in the context of clinical and patient interpretive practices.

Related to these ideas, a further shift in perspective is the idea that these body-tracking systems "see" differently from the ways in which a clinician sees. Much of the challenge with the development of these systems for use in clinical assessment can be helped in the acknowledgment of this concern. One of the key areas of design for these systems then pertains to the ways in which particular forms of feedback and representation of movement can enable clinicians to relate the ways they see and inspect movement in context with the ways that a system can see, parse, and infer movement patterns.

Moving forward from here, it is clear that we need to take these broader perspectives with respect to other forms of emerging ubiquitous sensing technologies that can have a role to play in new forms of clinical assessment practices. Body worn accelerometers, for example, will have different performance characteristics, benefits, and limitations for which similar kinds of dependency articulations need to be made with respect to clinical settings. We also need to consider these properties across a broader range of clinical conditions and any unique movement symptoms they may exhibit.

Perhaps more significantly, however, is to look beyond the more immediate paradigms of movement assessment practices. In much of our discussion here, the discussion of these systems has been in the context of current assessment procedures and environments. That is, a patient is still required to come into a hospital or clinic for the explicit purpose of being assessed and diagnosed. Yet much of the hope of these systems lies in their potential to shift the contexts of assessments in both spatial and temporal terms. For example, with the reduced cost and mobility of different forms of ubiquitous sensing there is an ever increasing possibility for movement assessments to take place beyond the clinic—in the home or out and about in the context of everyday living. Rather than intermittent and infrequent assessment, this starts to introduce future opportunities for on-going monitoring and real time assessment of movement symptoms. This in turn will enable the real time management of these. For example, Stone and Skubic (2013) installed a Kinect sensor above the door or a house to continuously monitor activity so as to detect early signs of illness or change in fall risk. More recently, efforts have been being made to create smart homes in which passive sensing is used to support the health of those living in them. This could potentially enable older adults to live longer in their own homes (e.g., CASALA Living Lab; SPHERE). A further well-articulated example is that of freezing gate in Parkinson's which, when initiated, can cause bad falls in a patient. Yet, such symptoms can be reversed with appropriately timed interventions, for example through listening to rhythmic music. Body worn tri-axial accelerometers could be used to detect in real time the movements associated with freezing gate (e.g., Takač et al., 2013), which could then trigger the music to be played. What we have seen in this chapter, however, is that realizing such scenarios goes beyond just exploring the feasibility of recognizing movement data patterns of a freezing gate from a body worn sensor such as an accelerometer. Rather, what we hope to have illustrated are some of the broader ways that the form and characteristics of these technologies may relate to the contextual factors within in these scenarios.

CHAPTER   3

Self-Directed Rehabilitation and Care

3.1   INTRODUCTION

Rehabilitation programs for those with movement-based disabilities and chronic conditions are facing increasing challenges. Significant demographic shifts are happening with many countries around the world dealing with the pressures of an increasingly aging population. Such demographic changes mean that health services in these various countries are having to cope with increasing numbers of patients suffering from chronic conditions, such as stroke and pain, and movement-based disorders. The demands of clinically based rehabilitation programs that require sustained and clinically supervised exercise sessions over long periods of time means that current structuring of rehabilitation healthcare will face difficulties in dealing with a growing number of patients. This increased pressure (in terms of cost, time, and accessibility) on these health systems has led to a pressing need to rethink the organization of rehabilitation interventions with a particular focus on more self-directed rehabilitation (e.g., Bodenheimer et al., 2002; Hirsche et al., 2011; Du et al., 2011). This in turn has led to a reorganization of the healthcare system in various countries (WHO, 2002) to facilitate the shift in responsibility for care from the clinician to the individuals suffering from chronic conditions (e.g., arthritis, diabetes, chronic pain, etc.),[1] with the medical structure playing just a marginal supporting role.

As managing one's condition is not an easy task when the support is limited, technology is being targeted as a possible way to address the problem. Until recently, the possibilities that technology could offer for self-management in older adults and people suffering from chronic conditions were limited to education, cognitive behavioral therapy, and self-monitoring of physical and emotional states. However, with the emergence of low-cost body-tracking technology and with the encouraging results obtained through its use in clinically directed therapy (either hospital- or home-based), there is an increasing interest in understanding how this sensing technology could provide support in self-directed physical rehabilitation. Significantly here, with an emphasis on low cost and ubiquity in relation to body tracking, these emerging technologies are not just seen as a way to reduce the load on overburdened health services but also as a way to provide new oppor-

---

[1]   http://personcentredcare.health.org.uk/person-centred-care/self-management-support?g-clid=CI2y-9PhjMQCFWIUwwodqEcAIw

tunities to make physical rehabilitation more effective. By shifting the emphasis from expensive, accuracy and precision in body tracking to one that is low cost, *sufficiently* accurate and widely available in patient homes, these emerging body-tracking technologies make rehabilitation sessions more promptly accessible to people since they do not have to travel to a hospital or to a drop-in hospital-supported gym. There are arguments too that patients have a preference for home-based care (e.g., Grönvall and Lundberg, 2014). In addition, as technology becomes wearable, it also opens new opportunities for offering a form of rehabilitation that is constantly available beyond physical rehabilitation sessions to encompass everyday functioning indoors and outdoors. Furthermore, by continuously tracking people's activity and measuring their physical and emotional capabilities, the technology could provide more personalized and effective therapy given the influence of individual factors (e.g., emotional, personality) on rehabilitation.

While the potential offered by body-tracking technology is very promising, it also opens new challenges that are not just technological but also methodological. The shift from clinic to self-directed home context raises important questions in relation to the ways that we think about the design of body-tracking technologies in rehabilitative interventions. Currently, an understanding of these factors has been underplayed relative to larger technological motivations driving innovation. As such, there is a relatively limited understanding of the constraints and enablers to self-directed home-based physical rehabilitation and thus how technology could address or support those (Singh et al., 2014). More recently, though, we are starting to see more effort to understand some of these factors and the implications these have on how we emphasize particular sensing characteristics in design (e.g., Tang et al. 2014b; Uzor and Baillie, 2013, 2014; Balaam et al., 2011; Grönvall and Lundberg, 2014; Alankus et al., 2010; Axelrod et al., 2009).

A key concern with these research efforts is with *motivation* and *adherence*. In part, this has related to inherent motivational barriers of repetitive and boring exercise programs and has prompted research to consider the role of body-tracking-based exergaming interventions in rehabilitation for movement disorders (Barry et al. 2014; Uzor and Baillie, 2013, 2014; Gerling et al., 2012; Mouawad et al., 2011; Alankus et al., 2010; Smeddinck et al., 2015). The key claim here is that the gamification of exercises provide additional motivation to adhere to a program in the absence of encouragement by a physiotherapist.

However, motivational concerns in self-directed rehabilitation are only part of the issues present in a home-based context. Indeed, it is crucial to recognize other features of the home-based context that influence how we might approach the design of body-based tracking systems for different forms of movement rehabilitation. For example, in contrast to the conduct of rehabilitation activity in dedicated spaces and times of the clinical context, there is a greater pressing concern for the patient at home to enable and support everyday functioning. This contextual concern takes priority in the configuration of self-directed rehabilitation over and above a more focused exercise aimed at greater physical progress.

Related to the everyday functioning concern, rehabilitation in the home implicates a wider social network of the family that needs to be considered in the way that self-directed rehabilitation programs can be conducted. The introduction of any technology or new behavioral demands into the home will not just impact on the patient but also on the everyday functioning and social organization of the *extended user network* (e.g., Balaam et al. 2011; Axelrod et al., 2009; Grönvall and Lundberg, 2014). Patients may have particular social dependencies and responsibilities that impact on the ways that the performance of rehabilitation can be configured. They may need to be cared for or may need to also be carers themselves. Practical constraints on the sensor technology and the physical demands of the exercise include the limited space of the home. We saw in the previous chapter, for example, how spatial constraints of certain clinical settings had implications for accommodating the characteristics of different sensors for body tracking. Similarly, rehabilitation systems that introduce dependencies on shared resources within the home may all impact on the patient abilities to exercise when required and where required. For example, if we consider an exergame based on commercial gaming technology this would typically be wired up to the household television. Yet, being a household television that is a shared resource among the extended user network, there are potential tensions arising from the introduction of such a technology that would demand negotiation over use. Such social concerns are very real practical considerations in the ways we design such systems for everyday self-directed rehabilitation.

Finally, self-directed rehabilitation by definition is reliant upon certain autonomy within the home (Fitzpatrick et al., 2010). This is not just a question of motivating the patient but also of informing and understanding what might constitute correct and appropriate movement performance. As we will see, understanding what appropriate movement performance is in these contexts is highly contingent upon the every day circumstances and state of the patient. Such contingency can have important implications for the ways that we approach the design of body-tracking technologies for self-directed rehabilitation.

In this chapter, we will explore these issues further in the context of two domains in which self-directed physical rehabilitation can play a part in patient therapy. The first of these considers self-directed physical rehabilitation in chronic musculoskeletal pain (CMP) patients, where not only physical but also psychological factors have to be addressed for the rehabilitation to be effective. The second of these then considers work on body-tracking technology, which has been designed to help older adults maintain their physical capabilities despite the pressures of aging on cognitive and physical performance. In both of these domains, there are some unique and interesting clinical concerns that lead us to think about different ways of characterizing body tracking in terms of how we measure and also in terms of the outcomes of using these measures. In both domains we also look to highlight some of the more contextual needs and barriers that are complementary to the underlying clinical concerns. For each of these we first present the context of each condition followed by the aims of the rehabilitation program. We will then critically discuss how the technology has benefited each group and explore the limitations that subsequently emerged. We will conclude by discussing what lessons can be learned from each of the two case studies and show how these can be generalized to self-directed physical rehabilitation in general.

3.2 FACILITATING PHYSICAL ACTIVITY IN CHRONIC MUSCULOSKELETAL PAIN

Chronic pain (CP) is a strongly disabling condition that affects people at physical, psychological and social levels. This condition is unfortunately very prominent in the Western world, with the most common form of it, chronic low back pain, increasing at a fast pace. One in five adults experience chronic pain in their life (Breivik et al., 2006)and for many of them the condition is non-treatable. 40% of chronic pain patients experience severe pain and are not able to live a normal life and about 20% lose their job. The economic cost is estimated to be in the billions of dollars (Kerns et al., 2011) in terms of impact on healthcare systems and also in terms of more general factors relating to lost productivity (e.g., unemployment and disability benefits).

CP is defined as pain that persists past healing (around three months) after an injury or with no identified lesion or pathology (Turk and Rudy, 1987). While acute pain usually resolves as the injury heals, CP continues indefinitely through central nervous system changes (Legrain et al., 2009). These changes lead to amplification of pain signals and dysfunction of its inhibition (Tracey and Bushnell, 2009; Bushnell et al., 2013). An increasing amount of empirical evidence shows that psychological, behavioral, and social factors also strongly contribute to *"the perpetuation, if not the development, of pain and pain-related disabilities"* (Kerns et al., 2011; Gatchel et al., 2007).

Of significance in CP is that the pain itself serves no purpose, yet people are wired to respond to it as an alarm signal with protective and avoidance behavior that is wrongly thought of as beneficial. On the contrary, for people with CP it is very important to remain physically active despite the ongoing pain. Being active reduces weakening and stiffness, inhibits the neurophysiological mechanisms underlying the spread of pain, and maintains confidence in being physically capable. Remaining active also underpins achieving valued goals (Gatchel et al., 2007), very critical to the person's well being.

As with other forms or movement-based rehabilitation, CP rehabilitation therapy is generally supported by a multidisciplinary approach, involving multiple stakeholders (clinicians, physiotherapists, and psychologists) that can lead to clear improvements (Kerns et al., 2011). As discussed above, due to pressures on healthcare resources, the difficulty and cost (i.e., due to traveling distance) of joining limited points of support (e.g., drop-in gym sessions for CP) when available, and the cost of accessing private rehabilitation programs, self-management has become the primary form of therapy. Self-management is very challenging and shares with other chronic conditions techniques and moderating factors that affect adoption and adherence to a therapy (Mynatt et al.,

2010). Similar to other chronic conditions, the acceptance and understanding of one's condition are paramount to regain control over one's life. Learning the skills to cope with the condition is also very critical. For example, learning to recognize and challenge negative emotions and replace these with positive coping thoughts is very important.

The rehabilitation of CP highlights a number of additional considerations that challenge the straightforward idea of the moving body as objective concern. In contrast to some other movement disorders and chronic conditions, there are specific barriers for CP patients remaining physically active that have a bearing on how body movement is considered and how rehabilitation is approached. Such barriers point to more psychological considerations in the understanding and treatment of movement. A key example here is that the experience of CP conveys a certain *threat* to the patient (Crombez et al., 2012). This generates fear of, and catastrophic thinking about, movement and activity, bringing new challenges to maintaining an activity-based rehabilitation program. In addition to this, CP is found to alter sensory perception, increasing attentional bias toward body cues that signal or are perceived as threat (e.g., increased heart rate, muscle tension). This excessive attention to threat leads to further distress and increases avoidance behavior (Simons et al. 2014). Research studies exploring this in CP have provided evidence of a relationship between anxiety and fear of movement and altered muscle activation typical of guarding behavior (Geisser et al., 2004; Watson et al., 1997). Excessive muscle contraction interferes with movement execution leading to increased pain and reinforcement of the belief that movement is the cause of pain. Of note here is that with a reduction in fear and with increased self-efficacy, these abnormal muscle activity patterns can return to normal (Watson et al., 1997). Finally, altered respiration patterns are also typical of increased anxiety in CP. Changes in respiration induce a perturbation in the posture and this is compensated for by movements of the spine and hips (Hamaoui et al., 2002), possibly affecting balance capabilities. The situation is further complicated in the case of an altered *proprioceptive* system (Lee et al., 2010; Volpe et al., 2006), whereby one may have a distorted perception about the amount of movement performed and the configuration of one's body.

3.3   TECHNOLOGY FOR CHRONIC PAIN REHABILITATION

Much of the technology to support people with CP in self-managing physical activity has thus far focused on enabling access to information through web-based resources and smartphone apps (Rosser et al., 2011; Rosser and Eccleston, 2011). Most of theses systems (e.g., Habit Changer: Pain Reduction on iPhone) provide information about pain and its reduction. Others (e.g., Pocket Therapy on iPhone) offer more meditative support with routines for strengthening and relieving tension, and relaxation. Meditation and relaxation applications are also emerging, for example, that explore sound and biofeedback to facilitate introspection and anxiety reduction through a more mindful reading of one's body changes (Vidyarthi et al., 2012). For other applications (e.g.,

Chronic Pain Tracker[2] on iPhone; WebMD PainCoach) a key enabler lies in support for setting and monitoring goals, activities, mood, and pain levels, as well as providing assistance such as reminders for medication and medical appointments. While these applications tend not to exploit more advanced tracking technology to provide run-time psychological support during physical activity, they nevertheless point to a number of critical components to consider in the context of body-tracking-based CP rehabilitation.

More recent work looks to provide support and motivation for doing "appropriate" physical activity and more directly relates to the particular concerns of the book. These solutions build on the success of full-body computer games and the potential they offer in tracking and motivating movement during physical activity. Initial work on the use of body-tracking technology for physical rehabilitation in chronic pain (e.g., Jansen-Kosterink et al., 2013; Schönauer et al., 2011; Omelina et al., 2012) has taken inspiration from various other rehabilitation systems built for other motor conditions where the aim is to recover lost motor capabilities (e.g., stroke, post-injury). Technology that has inspired this initial work ranges from robotic haptic devices that help their users by compensating for their weaknesses (Pauletto and Hunt, 2006; Wellner et al., 2007; Rosati et al., 2013) to more lightweight low-cost body-tracking technology (e.g., Kinect, strap-based accelerometers) that have potential to be used in the home (e.g., Geurts et al., 2011; Alankus et al., 2010). With a foundation of gamification to bring fun into the activity, the approach adopted by this technology is to track movement or muscle activity patterns and reward physical progress measured against a specific movement model. Parameters and targets defining the exercise and the movement model are set by a physiotherapist to ensure that the exercise space fits the physical capabilities and progress of the individual. Kinematic measures such as amplitude and speed of movement, ability to hold a particular position, deviation from a typical trajectory and muscle strength are amongst the typical gauges used to assess progress. The fact that the exercises are tailored by a clinician adds a level of safety that facilitates adherence to the program. In addition, new products are developed with the possibility of distance monitoring, reinforcing the role of the clinician as the main person responsible for the therapy and the supervisory pair of eyes.

While the core functionalities of these corrective and gamified technologies offer some potential help to CP patients, the particular *movement-correction* model overlooks the specific needs and barriers of people with CP in the context of their rehabilitation. Highlighting these helps us attend to the nuances of movement disability in specific conditions and contexts and what they may mean for how we successfully integrate body-tracking technology into rehabilitation solutions.

First, steady physical recovery of lost motor capabilities in CP patients, is not the primary aim of the rehabilitation as it might be with other conditions. Given that the level of pain fluctuates daily or even during the day, steady physical progress is not expected. Consequently, a more significant aim for movement rehabilitation in CP patients is to overcome *protective behavior* and improve patient confidence in their movement in spite of the experienced pain. While protective behavior (Keefe and Block, 1982; Sullivan et al., 2006) exhibited in CP may share similarity with compensatory behaviors adopted by people with motor impairments, a crucial difference for CP lies in the manifestations of underlying emotional states rather than just physical limitations. Manifest movement disabilities in CP need addressing from both psychological and physical perspective. Indeed a simple corrective model of physical movement can itself induce increased anxiety and avoidance behavior, among CP patients (Singh et al., 2014). In addition, learning to pace the physical activity according to one's body's physical and psychological resources is very important to avoid exposure to increased pain and related negative emotional states. Increased pain during physical activity may increase sensitivity to pain and also have lasting consequences as any setback may result in the person being bed bound for weeks.

Second, in contrast to other conditions requiring physical rehabilitation, the pain of CP is not temporary and is not due to the effort that physical activity imposes. Rather, pain is a constant and does not end once the rehabilitation session ends. As such, if we consider the motivational drivers of exergame based rehabilitation interventions in other conditions, the fun and distraction elements used to overcome the monotony of rehabilitation programs may not apply so well to CP if movement threat, anxiety and lack of confidence are not also addressed (e.g., Gromala et al., 2011).

In developing body-tracking technology for self-directed rehabilitation it is important to ground design decisions in a deeper understanding of the motivations and barriers CP patients experience in trying to keep active. Strategies relating to existing CP physiotherapy are pertinent here (e.g., Singh et al., 2014). First of all, rather than exploiting distraction mechanisms, physiotherapists will encourage patients to discover their body's physiological responses during exercise and to normalise them rather than perceive them as a threat. This focus of attention and self-awareness of one's body are also used to understand one's body's limitations (e.g., *"turn your head until you start to feel you are stretching your muscles"*)—a critical, yet challenging phase on the journey of learning to self-manage physical activity. Physiotherapists in traditional CP physiotherapy sessions facilitate body awareness by getting the patient with CP to focus on pleasurable bodily sensations (e.g., counting the number of breaths rather than the time spent holding a stretch) during anxiety-inducing exercises.

Second, since the main aim of physical rehabilitation in CP is to regain confidence in movement before gaining physical capabilities, there is a need to understand (and measure) what constitutes progress in self-management. *Psychological* progress (e.g., confidence in movement) and self-management skills (e.g., the ability to set one's own targets) are considered as important as *physical* progress. This means that progress targets in CP rehabilitation need to be adaptively defined in response to the circumstances of bad days (very high pain) or after setbacks.

The emphasis on self-management in CP also leads to further contrasts with other movement disorders in terms of the orientation of physiotherapeutic intervention. As Singh et al. (2014), argued, physiotherapists dealing in the area of CP consider their role not primarily as the directors and monitors of bespoke exercise programs but rather as teachers of self-management skills. Their role is more to educate about chronic pain and to help patients learn to read their body cues (physical and psychological) and use those cues to tailor their activity to their own physical capabilities and psychological needs. Graded support was provided during physical rehabilitation sessions, with the physiotherapist gradually refraining from providing support, advice and even feedback—a trajectory that needs to be followed in any technological intervention. Developing and continually monitoring such self awareness, however, is cognitively and emotionally demanding for CP patients as is the need to plan everyday activity in advance to ensure that the limited resources are sufficient to accomplish them (Sergio et al., 2015). In this regard, there is good scope for body tracking and sensing technology to facilitate the CP patient in monitoring their performance and increasing their self-awareness.

A further important target often overlooked in assessing rehabilitative progress concerns the ability to transfer performance improvements of physical exercise to more *everyday functioning*. Psychological factors and setbacks may be of greater significance when considering any transfer of movement improvements to better everyday functioning. Functional activities may be perceived as more demanding due to less controlled settings, social pressures and cognitive demand. Gromala et al. (2012) discussed how demanding it is for people with CP to engage in social life and how this often leads to avoidance and isolation. Furthermore, many patients, subsequently experience further psychological setbacks (e.g., deeper demotivation and frustration) when salient progress in their physical exercises does not noticeably translate it into everyday functioning. In thinking about the use of body-tracking technologies in these contexts then, consideration should be given to the ways in which their form can facilitate the transfer from exercise to everyday movement functionality.

3.3.1   GO-WITH-THE-FLOW: SONIFICATION IN MOVEMENT REHABILITATION

In order to illustrate these let us consider an example system, *Go-With-The-Flow* (Singh et al., 2015), which aims to create a rehabilitative exercise environment for patients with CP. At the core of the system is a capability for body tracking but the significance of the system lies in its holistic psychological and physical approach in contrast to the standard movement analysis and correction model. In adopting this approach, it points to a different way of thinking about body tracking in movement-based rehabilitation by exploiting interactive representation of tracked body movements rather than its straightforward capacity to measure and analyze. As in the previous chapter, the issue of movement representation is raised as a key issue here. While the concerns of the previous chapter pertained to the visualization of tracked movement data, our discussion here aims to highlight what a different material form of movement data representation can offer and in particular how these relate to key considerations of the rehabilitative context.

The use of sonification in the representation of tracked exercise movements comes with various rationales. The first of these relates to neuroscience research highlighting how our brains continuously use the sounds produced by our bodies when acting on the environment as a means to understand the position and posture of the body in the world (e.g., Botvinick and Cohen, 1998; De Vignemont, et al., 2005; Wolpert and Ghahramani, 2000). For example, the sound received when tapping on a surface informs the brain about hand position, arm length, and force applied (Tajadura- Jiménez et al., 2012). In this sense the sensory data from sound is an important basis for how we are aware of our own bodies in action. Neuroscience further highlights tight links between auditory and the motor systems in the brain. For example, listening to rhythmic sounds can activate motor areas in the cortex (Bengtsson et al., 2009; Peretz and Zatorre, 2005). As a consequence of this rhythmic acoustic feedback can be used to facilitate movement-execution by linking movement to rhythm (Kenyon and Thaut, 2005). Given these relations, altering sound-based feedback about one's body movements can be used to change our perceptions about movement. Sound was also something that could be easily attended to without fixating on a display. Importantly, this would allow greater freedom of movement as well as the opportunity for a patient to keep their eyes closed. This feature, combined with the sound being designed to be aesthetically pleasing, was a key in supporting both mindfulness and relaxation (Cepeda et al., 2006; Vidyarthi et al., 2012) in the patient within the context of their rehabilitation.

Based on these arguments various studies have used sonification of movement to facilitate movement or to make it more pleasurable with many showing its effectiveness in motor learning (Effenberg, 2005; Effenberg, et al., 2011, 2007) in studies of sports training (Kleiman-Weiner and Berger, 2006; Dubus, 2012; Schaffert et al., 2010). Movement sonification techniques have also been tried in various domains of rehabilitation such as spinal chord injuries (Rosati et al., 2013; Wellner et al., 2007) but also in patients with Parkinson's disease (Young et al., 2014; Rodger et al., 2014). Sonification of electromyographic data during rehabilitation has been shown to be a useful feedback tool for both physiotherapist and the person performing the movement to use the sound as a guide toward a target movement (Pauletto and Hunt, 2006). PhysioSonic (Vogt et al., 2009) transforms 3D movement analysis of shoulder joint kinematics into audio feedback to correct posture or coordinate a therapeutic exercise for the shoulder joint. This type of sonification of joint kinematics has been identified as a useful approach to help coordination in patients with a lack of proprioception (Chez et al., 2000; Matsubara, et al., 2013), which, interestingly, is a condition often described in people suffering from lower back pain (Lee et al., 2010).

In designing the sonification experience of the *Go-With-the Flow* system there were a number of key considerations.

1. *Provide an enhanced and pleasurable perception of movement* especially when movements are restricted and when a movement is associated with high anxiety.

2. *Provide a clear sense of progress through the exercise* so as to increase self-efficacy.

3. *Provide a sense of achievement* using anchor points to signal milestones and targets achieved.

4. *Facilitate going-with-the-flow* by reducing cognitive demands of continuous monitoring by the patient.

5. *Increasing awareness of avoidance strategies* and encourage movement exploration without provoking anxiety or fear of movement.

6. *Encourage preparatory movements* through pleasurable sound phrases distinct from the sonification of the exercise itself.

7. Develop self-management skills and enable self-calibration of exercise fit with the person's physical and psychological capability as well as the current pain level.

The initial *Go-with-the-flow* system focused on simple reaching forward exercises that are typical of movements inducing anxiety in CPP patients. Of interest to our concerns is that various forms of body-tracking technology were explored as the basis for the system. In the first instance, as illustrated in Figure 3.1, a smartphone with an embedded gyro was attached to the trunk of a patient during a reaching forward exercise. The embedded gyro is used here to measure the degree of bending of the trunk and depending on the incoming movement data, provides auditory feedback at fixed intervals of increased stretching. In the second instance, a Kinect depth sensor was used as the basis for body tracking with the system. We discuss this system in more detail later.

While various types of sonification possibilities were explored in this work (e.g., see flat sound or wave sounds in Figure 3.1), the key aims of the work were in the provision and self-calibration of appropriate anchor points. Milestones here are an interesting way to think about movement and the way that it is tracked. In essence, body movement here is characterized by the achievement of outcome goals of movement. In this sense, these aspects of body tracking are not dependent upon a detailed specification of a particular movement. Such characterizations offer latitude in the actual performance of a movement so long as the milestones are achieved. This, however, has implications for the low-level sophistication of the body-tracking equipment required—in this case it was a low-cost gyro in a smartphone attached to the torso of the patient.

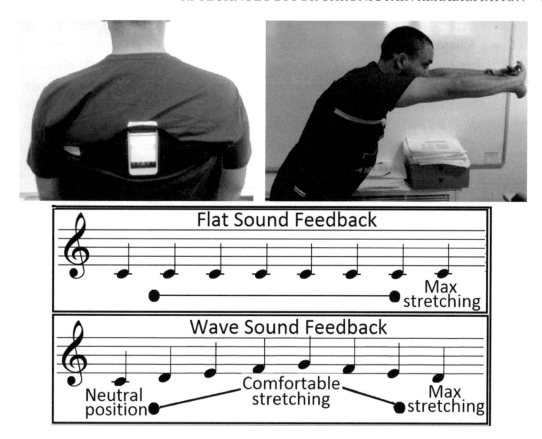

Figure 3.1: (Top) Device tied to participant's back and forward reach exercise. (Bottom) Sound feedback and exercise spaces.

Of further importance here is the particular approach to signaling and specifying milestones in this particular rehabilitative context. While employing sonified representations of body movement to signal milestone achievement or quality of movement have been well explored, what is particularly pertinent to these patients in these self-directed rehabilitative contexts, is a more specific way in which milestones are oriented to, each playing critically different roles and attending to both the physical and psychological needs of the patient. In this system, three different anchor points are used: a *start* position, a *comfortable* position, and a maximum stretch *target* position. Importantly, the patient calibrates these anchor points themselves and in so doing can be aligned with both their physical and psychological capabilities on that day (be it a good pain day or a bad pain day). This orients the system to one that reflects what the patient feels they can perform, which is a vital element in the threat-based associations of body movement when experiencing CP. In this way, the system does not set out to gradually increase physical capabilities but rather seeks to develop patient confidence in their own movement. Furthermore, the focus is not on the rights and wrongs of particular movement forms/targets or their subsequent correction. Indeed, a model of continuous improvement can easily lead both to overdoing activity as well as inducing negative feelings of demotivation (which are entirely in contrast to the needs of CP). On good pain days, the calibration points provide a sense of achievement (especially if the mood is low) by breaking down the activity into achievable parts of what the patient needs to do and motivates them to go a little further (Singh et al., 2014). Hence, the *comfortable point* rather than the *final point* may be seen as the milestone that facilitate not only maintenance but also steady and small improvements in physical and psychological skills over a longer period of rehabilitation. By not adopting the corrective approach of movement rehabilitation, the sonification of the Go-with-the-flow system facilitates the exploration and awareness of body movement. This is especially pertinent to the various protective behaviors adopted by CP patients to avoid pain (but which actually lead to greater longer term problems) many of which they are unaware of until someone points them out. The sonifications can provide cues about these different movement patterns that facilitate awareness in the patient and allow them to make a choice about how they adapt their behavior accordingly.

3.3.2   TRANSFERRING TO EVERYDAY FUNCTIONING: KINECT VS. WEARABLE SMARTPHONE AS A BODY-TRACKING DEVICE

As an alternative to the gyro-enabled smartphone approach in the initial Go-with the flow system, a Kinect sensor was also explored as the basis for body tracking in this context. What is of interest in these design alternatives is how they place in focus key trade-offs between body-tracking performance and other characteristics of the sensors in terms of their implications for everyday practice. The Kinect-based solution then provided more sophisticated movement-tracking capabilities that opened up opportunities for a wider range of exercises (reach forward, sit-to-stand) to be incorporated as well as also more creative sonification algorithms (Singh et al., 2015). Of note in the sit-to-stand exercise is that it can be facilitated by *preparatory movements* (i.e., bending the trunk forward while still sitting to provide momentum). People with CP typically avoid this preparatory phase of the movement for fear of increased pain and compensate for it by over-contracting the muscles and twisting the trunk and/or hand over a chair or legs as support. With the Kinect depth camera it is possible to detect and sonify the preparatory moment and the standing up movement separately enabling a richer therapeutic response to be created.

In spite of these more sophisticated opportunities opened up by the Kinect from a clinical perspective, these come at the expense of the benefits arising from the mobility and wearability of the smartphone-based solution to the body tracking. These benefits can be found in practicalities of achieving the desired patient transition from exercise to generalized gains in everyday function/ activities. With this in mind, the fact that the device could be used anywhere had enormous implications for self-directed rehabilitation at home for a variety of reasons. First of all, CP patients find it important to be able to interleave their exercises with other everyday activities that comprise everyday home life. Interleaving in this way offers greater opportunity to manage their time in relation to their other chores and family commitments in the household. The mobility allowed exercises to be more explicitly incorporated into everyday activity. For example, doing stretches when cooking or washing up. Patients would also be able to exploit the spatial and temporal flexibilities (in terms of when and where it is used) of the wearable sensor to augment their movement exercise at particular times and places; for example, performing stretches in bed when waking up to help them gently wake up and get started in the day. This also means that they can swap activities easily if they felt stiff after doing something for a long time. Being able to reconfigure these activities in terms of when and where they take place further enables the patients to organize their exercise so as not interfere with the routines and material resources with others in the household. In addition, the mobility and ubiquity allows them to configure their activity to avoid pain if they felt stiff after doing something for a long time.

Other ways in which the mobility of the sensor can be seen as facilitating the transition from exercise to everyday functioning was in ways that self-calibration could be related to the challenges of everyday household activities and targets—for example, the amount of stretching required to reach a book on the bookshelf or bending to load and unload the washing machine. Being mobile means providing the CP patients with a flexibility to relate to a wider range of their everyday challenges rather than being confined to using a Kinect that is more likely to stay in a single room. Over and above this, the mobility of the device offers opportunities for ongoing monitoring of their behaviors outside of the more prescribed household locations and schedules necessary with the Kinect. This can be of significance for example in helping patients become aware of any awkward movements or protective behaviors they habitually used while doing other everyday functional activity.

3.3.3   SELF-DIRECTED REHABILITATION AS PROCESS: FROM CLINICAL FACILITATION TO SELF-MANAGEMENT

The process of transitioning self-management skills from exercise to everyday functioning in the home is a gradual process and in designing these systems there needs to be a recognition of this. In the initial phases of the transition, the physiotherapist is very much present in what is clinically facilitated self-management. The role of the physiotherapist here is to help the CP patient learn how to be self-aware and eventually become self-directed. In this respect there is an important relationship between the environment and context of clinical facilitation to the environment and context of self-directed rehabilitation in the home. This casts an interesting light on the different body-tracking technologies in terms of their potential fit into this relationship. Indeed, both physiotherapists and patients look to relate the affordances of the various body-tracking technologies to the everyday practices bound up in this relationship. The mobile form factor of the smartphone is viewed as something that enables the device to be incorporated into initial sessions with the physiotherapist—to explore the setting of target points and identify body cues that could help them facilitate the setting when alone at home. Much of what we saw of the doctor-patient relationship in Chapter 2, such as the collaborative elements of this initial clinical facilitation context, are also at play here in terms of how we assess different affordances of the technology outside of their immediate body-tracking function. Part of this is about the ease of mobility back and forth between the clinic and home context that is largely enabled by the smartphone, and part of it is in the way that the device mobility facilitates the *in situ* conduct of social interaction between the physiotherapist and CP patient in these settings. Furthermore, as well as providing the basis for the physiotherapist to inform about appropriate calibration and appropriate use of the device, there is also the basis for the tracked movement data to be collected as evidence of progress, allowing the clinician to better understand people's behavior and how their strategies helped the person concerned.

3.3.4   TRACKING AFFECTIVE STATES AND PAIN LEVELS

Monitoring progress and exercise sessions allows the physiotherapist to judge whether they need to intervene further in the level of support needed. This judgment relies on assessing both the physical and the psychological levels of progress with both of these working together in tandem in the manifestations of patient symptoms. Intriguingly, as our tracking technologies are becoming more sophisticated, there are interesting moves to augment and even automate some of these judgments to enable even easier transfer from clinically facilitated rehabilitation to self-directed home rehabilitation. That is, these body-tracking technologies will not just provide a sense of physical movement capabilities/disabilities but also an important indication of the affective states associated with particular movements in patients. This has remained underexplored thus far in the design of rehabilitation technologies. Most of the work on automatic detection of pain under particular conditions of movement and posture has focused on facial expressions. This is partially due to the fact that the work in automatic recognition of facial expressions is much more mature than that for the other modalities as based on established models from the psychology literature (e.g., the Facial Action Coding System (FACS; Ekman and Friesen, 1978). Using computer-vision methods, various studies have investigated the possibility of automatically recognizing the level of pain expressed with reasonable success for binary expressions (pain vs. no pain) (Lucy et al., 2011), but with decreasing performance when needing to distinguish different levels of pain (Hammal and Cohn, 2012; Kaltwang et al., 2012; Romera-Paredes et al., 2012, 2013).

While facial expressions relating to pain are an important communicative channel, they do not provide a full understanding of how a person feels in musculoskeletal chronic pain in and of themselves. As discussed earlier, further insights about a patient's affective state in relation to pain and the level of their confidence in functioning can be found in the use of protective behavior during the execution of a movement. Understanding and improving awareness of these protective behaviors is not just important to assess affective states but is also important in improving them. Protective behaviors are symptomatic but also have social consequences. Patients exhibiting these behaviors are often perceived negatively by others, being judged as less likable and less dependable (Martel et al., 2012; Ashton-James et al., 2014). This negative perception can lead to social exclusion with subsequent negative impact on the patient's mood. Protective behaviors also occur often without further indication from facial expressions of pain, which highlights the importance of tracking these behaviors with a view to understanding affective response in the context of self-directed rehabilitation (Aung et al., 2013). Recent research is demonstrating the feasibility of using body-sensing technology and machine-learning techniques to automatically recognize and classify these protective behaviors exhibited in the context of a particular rehabilitative exercise (Aung et al., 2013, 2014, 2015). This points to a really intriguing way in which these body-tracking technologies can bring both the physical and critically the psychological components of self-directed CP rehabilitation together. In detecting how a person is feeling through their exhibited movement behaviors, opportunities are opened up to provide appropriately personalized psychological support to ensure adherence to a program of physical activity and progress.

3.4   EXERGAMING AND BALANCE REHABILITATION IN OLDER ADULTS

Following on from our discussion of body-tracking technology in the self-directed rehabilitation of chronic pain, we move on now to consider a different domain of rehabilitation, namely that of balance rehabilitation. Again, our concerns here are, in part, with the unique clinical characteristics of the disability and the implications this has for particular forms of body-tracking approaches. But our interests also lie in how the particular population characteristics and residential contexts associated with this clinical domain impact on design considerations for these technologies. More specifically, we want to consider the use of body-tracking technology in relation to the rehabilitation of older patients who experience balance difficulties and consequently fall risk. With many of these patients also living in residential care as opposed to living in their own home, this domain offers us the additional opportunity to articulate additional contextual factors that impact on the design of body-tracking exergame balance rehabilitation.

3.4.1   BALANCE AND FALL RISK IN OLDER ADULTS

Maintaining our balance is a complex process that requires sensory inputs from the visual, vestibular (a mechanical fluid system located in the inner ear) and proprioceptive (body position) sensory systems. All these systems work together to give us our sense of balance. In order to stay standing upright, a person must continually monitor the convergent information coming from these different sensory systems and perform minute adjustments to the position of their body to keep the body's center of mass in a balanced position. Good balance control is fundamental to locomotion and forms the basis on which other movements are built. For instance, reaching for a book on a shelf or getting up from a chair not only involves the movement of limbs but will also result in a change in the position of the center of mass as the limbs move. This requires that continual postural adjustments be made to ensure that the person can successfully complete the action without falling over (Winter, 1995).

In older adults, however, the perceptual systems in the brain which underpin our balance control do not appear to integrate this sensory information as efficiently, making it more difficult for them to adapt to situations that require postural adjustments (Lord et al., 1994; Westlake et al., 2007). Postural control is further compromised when an unexpected perturbation, such as stumbling, occurs which requires rapid adjustments in movement in order to prevent a fall (Thelen et al., 2000). These age related changes in the body negatively impact on both functional balance and general physical stability leading to a significant rise in the number of falls in older adults (Alexander, 1994; Maki et al., 1991, 1994; Laughton et al., 2003).

This deterioration in balance performance coupled with an increased incidence of falls has catastrophic consequences on the lives of older adults. Recent statistics from the U.S. show that one out of three adults aged 65 and over falls every year. More troublesome, falls in this age group are the leading cause of injuries and hospital admissions for trauma. In 2008, nearly 20,000 older adults actually died from fall injuries while in 2009, emergency departments treated 2.2 million non-fatal fall injuries among older people with more than 581,000 of these being hospitalized at an estimated cost of $28 billion (Centers for Disease Control and Prevention, 2013). The impact of falls and falls related injuries is far-reaching, often negatively affecting general mobility and independence. Interestingly, older adults who have experienced a fall often develop a fear of falling which can further add to the negative effect the fall has had on quality of life. Although older adults who have experienced a fall will often choose to limit their mobility to reduce the perceived risk and fear of falling, this in turn reduces engagement in physical activities, which in fact accentuates the deterioration in balance control, leading to a further real increase in falls risk.

To break this vicious circle between increased risk of falls and the debilitating effect of fear of falling many healthcare professionals often educate older adults about how to build confidence and prevent falls from happening. These include things like the "Stay on Your Feet" (https://www.stay-onyourfeet.com.au) program in Australia that suggests that the incidence of falls can be reduced by following advice such as: be active, manage your medicines, walk tall, check your eyesight regularly and improve your balance. However, education and advice are only part of the solution here and there remain significant challenges with following these recommendations in the course of everyday practice. Few of the educational initiatives and associated training exercise programs actually have specific balance control training exercises that can be easily implemented in a person's own home or place of residence. While exercise classes such as Tai Chi, Pilates, and Yoga are recommended for their ability to improve balance control, access to these is not always easy for the older adult, and is often deemed one step too far for those who need the most help.

One type of intervention that has been found some success in improving the balance control capabilities of older adults is a balance-training program. In such programs, older adults are guided through a series of specially designed movement exercises that focus on how to improve balance control with a particular emphasis on the sensed position of the body in space (proprioception). A recent study (Barnett et al., 2003) showed that when physiotherapists used this balance re-training approach with balance-challenged older adults, the adults ended up less likely to fall relative to those patients where the intervention focused only on educating older adults about fall prevention. That is, the use of balance control exercises can help individuals train their central nervous system to respond to the different sensory challenges allowing them to balance better and therefore reduce the risk of falls. As with other forms of rehabilitation, however, the clinical benefits of such programs are often diminished due to poor uptake and depleted motivation to continue with the program long term (e.g., Rose, 2008; Nyman and Victor, 2011; Uzor and Baillee, 2013). Indeed, by not adhering to recommended levels of exercise or even discontinuing completely, there can be a resulting decline in any benefits of any exercise that has already been done—leading to an increased risk of falling (Sherrington et al., 2011).

Furthermore, and in line with the arguments set out in the beginning of the chapter, the practical and economic viability of current healthcare configurations which require a physiotherapist to pair with an older adult patient, are not sustainable in an increasingly older population. Again, this creates an impetus to consider alternative configurations of such healthcare that shifts some of the responsibility back toward the person suffering the condition.

3.4.2   BODY-TRACKING TECHNOLOGY FOR BALANCE TRAINING

As with other areas of self-directed movement rehabilitation, balance and fall prevention interventions have turned to *exergaming* as a potential solution to the current challenges. Following other areas, such as stroke rehabilitation (e.g., Geurts et al., 2011; Alankus et al., 2010), the key driving factors relate to the gamification arguments whereby the characteristics of play help overcome the intrinsic monotony of rehabilitative exercise. By making these activities more fun and also more challenging (e.g., Levac et al., 2012) with respect to the specific needs of the patient, there is an argument that such *exergames*, with tailored targets and challenges will enable better adherence to a long-term rehabilitation program (e.g., Uzor et al., 2012). Very importantly, by using body-tracking technology in a game real movement can be directly mapped onto the game environment, offering huge potential as a low-cost movement-based training system where the design of the games can be specifically targeted to enhance specific levels of motor control in rehabilitation contexts. These movements are given additional layers of meaning and purpose outside of their primary rehabilitative function. Other arguments pertain to the ways exercises are presented to the older adults for them to follow. For instance, in traditional balance therapy contexts, the presentation of the exercises themselves may be in booklet form. While these offer a certain simplicity and convenience to be used wherever, they also do little to articulate some of the appropriate dynamic qualities of movement that are part of their correct performance. The use of body-tracking technology in rehabilitative balance exergaming contexts provides the basis for informing about these qualities of movement for better therapeutic effect. For example, Uzor and Baillee (2014) discussed how dynamic qualities such as pacing of movement can be represented in dynamic visual form and related to the tracked movement of the older adult. This offers a much richer way of guiding the correct movement.

The low-cost body-tracking technologies considered in this space have included the commercial games controllers notably the Wiimote and the Kinect. While both of these technologies have a certain appeal in this space there are also some challenges associated with their use. The Wiimote, for example, has been demonstrated in the context of rehabiltation (Alankus et al., 2010) but its use in this context of older adults has been criticized on the basis of them needing to be strapped to the arms of the patient (Uzor and Baillee, 2013) which is uncomfortable and potentially constraining freedom of motion. These authors also suggest too that some of these drawbacks could be overcome with the use of depth sensing camera such as the Kinect. The obvious appeal of the Kinect here is that it allows rich motion tracking without encumbering their movement and without the need to attach sensors to the body. However, as Uzor and Baillee (2013) highlight, older adults and, in particular, those with balance-related movement disability, present a particular challenge to the use of Kinect. Related to some of the arguments in the assessment of multiple sclerosis patients discussed previously, the balance-impaired older adults are dependent upon various forms of support surfaces or walking artifacts (walking canes and Zimmer frames) to prevent falling during everyday activity. These artifacts ultimately act as an extension of the patient's body, which subsequently interferes with the algorithmic assumptions about the body bound up in the computer-vision algorithms underpinning the Kinect skeletal tracking. As such, in spite of the richness of motion tracking available with the Kinect, these do not account for some of the everyday practicalities of movement in older adults, making it less suitable as an option.

In light of these challenges a number of alternative technologies have been considered to better address the specific needs of this context. Uzor and Baillee's (2013, 2014) work, for example, opted for small purpose-built inertial sensors (accelerometer, gyroscope, and magnetometer) to be worn on the relevant parts of the body. Such solutions, however, are not without their own challenges in relation to the older adult target group. The adults in these circumstances did not want to have wearable sleeves with in-sewn sensors and did not want to be forced to modify their clothes in order to accommodate everyday use of these inertial sensors. It was also important for the sensors not to be stuck onto their clothes in any way. As a consequence, velcro-straps were used to allow the sensors to remain in place as well as allowing them to be easily taken on and off. While there was a desire for small and unobtrusive sensor form factors that were inconspicuous and which facilitated freedom of movement, this needed to be traded off against the practicalities of self-managed application of the sensors in everyday rehabilitation. Patients of this age may have other physical or cognitive impairments (e.g., impaired vision, arthritis, etc.) that need to be additionally accounted for here. The form factor of the wearable inertial sensors, then, also needed to be large enough to be seen, correctly identified, correctly attached and easily removed. In working with elderly patients, Uzor and Baillee (2013) highlighted some initial difficulties experienced by these users such as placing the inertial devices upside down. While these were often quickly overcome with experience, the broader significance of the work highlights usability factors that extend beyond the *centrality of body measurement* but which need to be carefully considered when deigning these systems for independent everyday use by this specific user population.

A second alternative takes a different tack and looks to *balance platforms* as the basis for these rehabilitative balance-training interventions (Betker et al., 2007; Sihvonen et al., 2004; Wolf et al., 1997). What is intriguing about balance platforms is that they highlight a different way of thinking about body movement when compared with the more obvious characterizations present in full motion capture systems. Rather than being concerned with the rich measurement of moving limbs, the balance platform can characterize the motion of the whole body in a single *Center of Pressure* (CoP) measure—a direct measure of controlling balance control (Winter, 1995). Of significance with this measure is that a particular balance position, or center of pressure, can be obtained through an infinite number of limb position configurations, which makes full-body motion detection a less optimal means of measuring, or indeed calculating balance, as a controller in an exergaming game context. As the CoP is the result of full-body movement, it is best to use the *outcome* of the movement rather than the physical means of movement to control balance in a game (Wulf and Prinz, 2001). More specifically, body-tracking technology, such as balance platforms, allows the individual to become more aware of their own body position, a critical factor in balance training, while still becoming immersed in the narrative of an exergame.

In recent years, the feasibility of using balance platforms from commercially available gaming technology has been explored in the context of rehabilitative balance training for older adults. The most studied system is the Nintendo Wii-Fit balance board (Wii-Fit), which has been evaluated in this context by several researchers (e.g., Pigford and Andrews, 2010; Williams et al., 2011; Bainbridge et al., 2011). For example, Pigford and Andrews (2010) carried out a study in which an 87-year-old man with a history of multiple falls underwent 2 weeks of Wii balance training coupled with standard physical therapy. The outcome of this intervention was a significant improvement in *functional balance* and *balance confidence* as well as an increased gait speed. Likewise, Williams et al. (2011) demonstrated significant improvements in functional balance, as measured by the Berg Balance Scale (Berg et al., 1992) in 22 community-dwelling healthy, older adults, following 4 weeks of Wii Fit game play. Similarly, Bainbridge et al. (2011) looked at a small group of 8 community-dwelling older adults with perceptible declines in balance and found clinically significant improvements on the Berg Balance Scale following a 6-week intervention program using the Nintendo Wii Balance Board.

3.4.3   DESIGNING A BALANCE TRAINING GAME

While these studies showed some initially promising results, they utilized off-the-shelf commercially available games for the Wii-fit. Such games are targeted at a much younger audience, with a core identity of weight loss and fitness, and are therefore not very sympathetic to the restricted movement capabilities (e.g., range of movement and reaction speeds etc.), potential cognitive impairments and needs of an older user population (Laver et al., 2011; Smeddinck et al., 2012; Gerling and Masuch, 2011). Indeed, playing some of the off-the shelf sports games would likely pose a falls risk to an older user. For this technology to work in a rehabilitation context it is important that the gaming interface is adapted to the action capabilities of the user and the training needs are mapped onto the game content (e.g., Awad et al., 2014; Young et al., 2011; Clark et al., 2010). With this in mind the rehabilitative games need to be designed in such a way that they encourage a stretched range of movement but that are also sensitive to the potential limitations of older adults. This means designing a balance training game that uses the levels of game play to progressively increase the quantity of CoP displacement required to successfully play the game.

At the heart of these rehabilitative interventions are concerns with motivation, challenge and engagement as well as a salient sense of patient progress. Such progress here is both in terms of improved physical capabilities but also levels of confidence in performing a wider range of movements at greater speeds. In this sense an appropriately designed exergame in this space will adapt to the ongoing capabilities of the patient, making balance movements progressively harder with faster and larger movement. Task performance feedback is an essential component in these systems in addressing the motivational and affective factors relating to movement and balance. Such performance feedback becomes a source of competition with oneself over the course of a rehabilitation program but also as a potential source of competition with others—providing additional motivational drivers to increase adherence.

Along the lines of the above, Young et al. (2011) developed a balance training exergame system utilizing a specially modified Nintendo Wii balance board. This modification of the balance board was of note because it highlights a rationale behind the sensor choice in this context that pertains to the critique of Kinect discussed earlier. The balance training system was comprised of a commercially available Nintendo Wii balance board, surround foam platform with plywood base and a Zimmer frame embedded in the foam platform (see Figure 3.2).

Figure 3.2: The modified Wii Balance Board.

This custom-made support frame and safety mat were incorporated as an additional safety feature to reassure participants who may have felt unsteady stepping onto the Wii balance board on its own. As safety features they are important in and of themselves but there is an important relationship here with the psychological underpinnings of balance rehabilitation. The very presence of these safety paraphernalia integrated into the body-tracking system provides the older adults with a level of reassurance. What is significant about this is that it gives the users the confidence to explore the capabilities of their bodies without the fear that it is detrimental to their balance rehabilitation.

Shifting weight from one foot to the other causes the center of pressure to move dynamically from one side to the other. The CoP data from the balance board is directly linked to the in game visuals generated by 3D graphics software. By using this type of interface, any changes in body position, and subsequently postural control, will directly influence what happens in the game. This link between what the user perceives on the screen and how they move will form the basis of the training and feedback. Task constraints (the way the visual stimuli move) will be manipulated in such a way that the user will be required to move their body to a greater or lesser extent in certain directions, with the increasing level of difficulty of the balance training program being controlled in a progressive way.

Let us consider two games that were designed specifically as a part of this balance training system, namely *Apple Catch* and *Bubble Pop*, which differ in both the level of complexity and required movements to play the game (e.g., side to side and front back postural control) (see Figure 3.3). In *Apple Catch*, CoP data is used to control the side-to-side movement of an onscreen basket in order to catch apples falling from a tree. The position of the apples on the tree and the speed with which they fell could be varied to increase/decrease the level of difficulty. In *Bubble Pop* the player was required to control their balance to move an onscreen lobster up and down and side to side so that rising bubbles could be popped with its body. Complexity was added by modifying the games slightly so that the appearance of a certain cue, such as a shark, would trigger a different action response. For instance, when a shark appeared the gamer had to stop popping bubbles, find a rock and hide behind it until the shark left the screen (*Avoid the Shark*) (see Figure 3.3). Here the aim of the exergame design was to train the patient to switch attention quickly from one task to another—a key component when performing everyday tasks that can often result in compromised balance responses in older adults. By incorporating attention switching into the game, this important component of postural control could be trained. The other game variation (*Smart Shrimp*) (see Figure 3.3) involved dual tasking, namely performing two tasks at once. In this case the gamer had to solve a simple cognitive based task (e.g., add two numbers together or find the missing letter in a word) and then search the rising bubbles and pop the one that contained the right answer.

Apple Catch

Bubble Pop

Avoid the Shark

Smart Shrimp

Figure 3.3: Screenshots from the range of games presented to participants: Apple Catch, Bubble Pop, Avoid the Shark, and Smart Shrimp.

3.4.4   UNDERSTANDING REHABILITATIVE GAME USE

The ways that we can understand the use of these body-tracking exergames in rehabilitation is an important concern for us. In part, some of the understanding can be found in classic quantifiable and controlled measures of performance outcomes. If we consider the above balance training games (*Apple Catch*, etc.) in terms of the performance outcomes, a study by Whyatt and colleagues of these games being played over a 5-week period (at least 20 min game play on 10 different occasions) showed significant improvements in balance performance among older adults (Whyatt et al., 2015). This was both in terms of (a) *functional balance* as measured using the Berg Balance Scale (BBS) (Berg et al., 1992)—a test that looks at how well someone can turn around, stand from sitting, pick up an object off the floor) and (b) *balance confidence* as measured using the Activities Balance Confidence (ABC) Scale (Powell and Myers, 1995)—a test that assesses how confident someone feels performing activities of daily living such as walking to the shops or taking a shower. The BBS is scored by a trained assessor while the ABC scale reflected the participant's own subjective rating of their confidence when performing certain activities of daily living that could be perceived as posing a falls risk. Objective measures of balance performance that used the Balance Board and CoP based on screen control tasks also provided further quantified evidence of balance improvements as a result of playing these balance-based games.

While these kinds of quantified characterizations of rehabilitation outcomes are an important part of our understanding, it is key that we further consider other aspects of how these exergames may function, and work in everyday social practices of older adults. An interesting contextual component relating to this particular population of older adult users is that a significant proportion of them will be based in shared residential complexes with other older adults and carers. Such settings will bring with them unique social and contextual dynamics and practices that may play out in both positive and negative ways on how such exergaming technologies can be effectively integrated (e.g., Harley et al., 2010; Ulbrecht et al., 2012; Gerling et al., 2015).

Generally speaking, the social context of the residential care home is considered to be an important factor in facilitating engagement with exergames in ways that might contribute to ongoing motivation to participate (e.g., Marston et al., 2013; Jung et al., 2009). Playing games together, for example, introduces a sense of competition that can be a potential driver to ongoing adherence in rehabilitation programs. This phenomenon also was apparent in studies of the above balance training games, where the sense of competition among the care home residents was something that was a topic of conversation between the residents (Whyatt et al., 2015). It was clear from the conversations that the social competition here was something they felt spurred them on to continually improve their performance relative to their own previous scores and relative to the other participants. While there is evidence to suggest a motivational source in this kind of residential setting, other related research in this space suggest that the contextual circumstances of these kinds of effects may be further nuanced. In the study of Gerling et al. (2015), for example, the observed support for these social factors in motivating the older adults was dependent upon the particular care home and types of older residents. Gerling et al.'s work also revealed some negative effects of competition in exergame play among more vulnerable older adults (which in turn introduced frustration and discouraged these adults from continuing their participation). The adults in question became increasingly self-conscious about their relative performance levels. This led to a decrease in confidence among these residents, which is ultimately detrimental to any exergame intervention where enhancing confidence is a key factor. These begin to further highlight some of the key social and affective foundations of body-tracking-based rehabilitation that need to be considered in design and use.

A similar story can be found elsewhere when considering other aspects of the social setting on participation. In Young et al.'s (2011) balance training games (*Apple Catch*, etc.), there was an explicit attempt to design the system to foster a social context of rehabilitation play rather than not just focusing on a single player. More specifically, they included a large screen on the system so that it would be viewable by an audience rather than just the player on the balance board. Putting the game with a large screen in a community room enabled the game play to become a social event.

There was evidence that such a social context drew players out of isolation, and provided spectator support for players. This was seen to improve the well-being and confidence levels of certain adults in the care setting that could encourage ongoing participation and adherence. Again, though, there are certain contingencies that can also operate here as seen in Gerling et al.'s (2015) study. So, for example, the experience of being watched playing games was intimidating for certain more vulnerable players. Given the importance of confidence building in successful balance retraining in these kinds of exergame balance rehabilitation programs, some of these inhibitory effects may be detrimental to the patients (as well as the more positive social effects) are worthy of further consideration in the ways we think about their design and use.

When used as a shared resource in a public setting we can see how this may afford particular forms of community building that are linked to rehabilitative success. But being a shared resource can also bring about challenges in the practical integration of the rehabilitation games into every day routines. Again, Gerling et al.'s (2015) work, which used the Kinect and was not designed to train balance, is informative here highlighting various tensions and frustrations (as well as the positive outcomes) among residents using exergames about having to share the resources and getting cross with other residents about slow handover times for the game. These challenging social dynamics may be exacerbated by features of the games and tracking systems being used. The standard Kinect games, for example, are more difficult to use for many older adults (than, say, the above bespoke balance training games) and many also do not find it intuitive how to control the game. The social context in this sense is likely to amplify the frustrations and potentially negative responses to particular systems which further highlights the need for bespoke design of theses systems for the specific context and population. While the exact nature of these reactions is bound up in the particular *in situ* circumstances of these encounters, they nevertheless highlight some of the design issues arising when positioning these body-tracking-based technologies in the context of every day practice with population of users.

Furthermore, practical issues with a significant bearing on the design and use of these body-tracking technologies in practices of the care home concern the extended user network of the residential care home. In particular, among this population of players, the idea of self-directed rehabilitation is not quite so straightforward. There are various dependencies on the availability of staff and volunteers who may be needed to offer support to elderly residents in setting up the exergame sessions (e.g., Gerling et al., 2015; Marston et al., 2013). Depending on their particular cognitive and physical disabilities, some residents will require much stronger support from dedicated staff. As we discussed in the chapter on assessment, the introduction of a caregiver into the context has not just implications for which body-tracking technologies might be appropriate, but also introduces practical considerations affecting the successful integration of these rehabilitative technologies in residential care settings. For example, it may require the use of additional support staff to ensure that the opportunities remain open to residents. But it also points to the need for possible design solutions that look to alleviate the need for support, for example, by making it more suitable for independent use by this group.

3.5   CONCLUSION

In this chapter we discussed how body-tracking technologies can be used as the basis for input and control of rehabilitative exergames that can encourage people to, not only move more, but also move better. In other words, through the game design players are invited to perform certain types of movement to interact with the game system, which can improve movement training and help with rehabilitation. As highlighted in the work of Levin, et al. (2010), there are key design considerations at stake when developing games for rehabilitation purposes. In particular, here is the importance of exploring the unique capabilities of different clinical target groups to provide accessible game play, to create meaningful in-game metaphors that relate to clinically significant real world movement and action, to support adaptability of the game to include a broad range of players and gaming situations.

In presenting the two case studies of chronic pain rehabilitation and balance training, we have had the opportunity to highlight some of the enormous potential of body-tracking technologies for rehabilitation in general. In looking across both domains we have been able to point to common concerns across these and other forms of rehabilitation to which such technologies might be applied. In particular, what is apparent from these case studies is how we need to look beyond the body as an objective and physical concern. Indeed, what is key to these applications is a combination of physical, clinical, motivational, affective, economic, social, and practical factors that need to be understood in terms of how we design for new scenarios and in terms of how we trade-off the characteristics of different forms of body-tracking sensors.

Of particular significance are the economic drivers at the heart of much of these new rehabilitative initiatives. With the pressures on health systems around the world, the significance of low-cost body tracking vs. more sophisticated high-end systems lies in their potential to completely reconfigure the ways that rehabilitation programs are delivered. Low cost and ubiquity of sensors are the significant features of the new sensors enabling the important shift from clinically directed to self-directed movement-based rehabilitation.

A further feature of these domains is that they introduce new ways for us to think about the different ways that we can track body movement and the level precision required. For example, in rehabilitation a key goal is to promote particular forms of movement rather than simply to measure it per se. For example, one can encourage a stretch movement by placing a target high up without actually prescribing the movement trajectory necessary. This is not to say that more rich tracking is not important in these domains but rather that there is a richer repertoire of tracking opportunities and capabilities that relate differently to these goals than what is needed in assessment for example.

The expansion of body-tracking ideas from measurement to interactive control in these scenarios is of note here.

Tracking movement in these contexts is important to provide feedback about physical movement progress and to allow the patient specific tailoring of activity to meet their needs While still being sympathetic to their action capabilities. However, for this technology to be effective, critical aspects of self-management need to be taken into account. By exploring across these two case studies we can see how important it is for the technology to be based on an in-depth understanding of the self-management program aims and contexts for each specific condition. Barriers and needs must be identified and understood in the context of the condition with appropriate technology being used to implement the program.

One thing emerging from these studies is the significance of emotional and effective factors that often are overlooked in more objective treatments of body-tracking technologies. Such factors are intrinsically bound up with physical movement is a range of ways. For example, if performing certain types of movement is perceived as being dangerous, this may result in protective or avoidance movement behaviors, which may be detrimental to the person. Failure to take these into account in design and deployment of these technologies may hinder adherence to the program of physical activity. But what is of significance here is how it turning attention to these opens up new opportunities for innovation with body-tracking technologies. The chronic pain study illustrated how cheap sensing technology can be used to increase awareness of one's own physical capabilities through tracking and representing movement-based information that is easy and pleasurable to attend to and devoid of information that may instead lead to negative thoughts (Kerns et al., 2011).

Everyday social context was also an important consideration in this domain with the two different case studies illustrating how this is manifest in a variety of ways that have implications for our understanding of these technologies. We see evidence of social facilitation that can enhance well-being motivation and adherence but also evidence in vulnerable adults of some inhibitory aspects of social context. Having some literacy of these various factors and social dynamics is critical in the design choices that we make with respect to the body-tracking sensors and the broader systems to which they are attached. But the social context too is vital to consider in the ways that rehabilitation practices and outcomes are integrated into everyday life and everyday functioning. In part, this is about considering the practices and demands of an extended user network in these specific settings with technology being of a form that allows patients to attend to other domestic commitments. But it is also important to ensure that there is a transfer from the physical activity in game based sessions to activities performed in real-life. Increasing self-efficacy in both physical activity sessions and in functional living tasks is the primary goal of self-directed rehabilitation in chronic conditions. Unfortunately, most of the work in the design of technology for physical rehabilitation has tended to focus only on facilitating physical progress rather than how it might transfer to everyday life. Attending to these issues emphasizes the benefits of particular sensor form factors over others, e.g., mobility of wearable sensors facilitates integration with everyday life while the fixity of Kinect arguably constrains integration. Looking at these technologies in these terms rather than from the centrality of body measurement perspective is key to the future development and success of these technologies.

CHAPTER 4

Interactions for Clinicians

4.1 INTRODUCTION

Over the course of several decades we have witnessed huge advances in medical imaging technologies that provide us with new ways of seeing inside the bodies of patients. With these new possibilities for viewing the patient body, we have seen in particular their widespread adoption within the operating room, transforming the ways that surgical procedures can be carried out. If we look around any modern-day operating room, it is quickly apparent that there are a large number of visual displays for accessing a wide range of imaging resources relating to the patient's anatomy and clinical condition. These may include, for example, computer tomography (CT), magnetic resonance imagery (MRI), ultrasonography, fluoroscopy, endoscopy, and various other procedure-specific imaging applications. Such medical imaging may be produced pre-operatively and subsequently made available for access in the operating room, or they may be produced intra-operatively during the conduct of the procedure itself. These varied images resources are created, assembled, and used during surgery for a wide variety of purposes such as real time guidance, reference, clinical discussion, diagnosis and for aspects of procedural documentation.

In order to understand the significance of these imaging technologies in the operating room, it is crucial that we extend our thinking about them beyond the idea that they are self-contained and self-explicating visual representations of data that simply allow a clinician to "see" inside the patient body and make visible what would otherwise be non-visible. Rather, as authors such as Lynch (1990a, 1900b) and Goodwin (1994, 2000) argue, what is crucial to our understanding is how these images and representations are situated in the particular practices of surgery and the part they play in the production of meaningful action within the operating room. Goodwin coined the term *Professional Vision* here to refer to that range of ways and actions through which clinically relevant features of these images and relationships between images are collaboratively made visible and interpretable by the surgical team as a whole. This includes the ways that these images are constructed, how they are assembled and viewed together, how they are dynamically manipulated, how they interacted with, how they are attended to, and how they are pointed at or gesticulated around in the context of talk and discussion with colleagues (cf. Alač, 2008).

This leads us to a number of key points. First of all, in the context of any surgical procedure there is not a singular prescribed set of image views and manipulations to be performed. Rather, the surgeons and clinicians make particular judgments about how to assemble and use the images based on in-the-moment contingencies of surgical context. All sorts of judgments and clinical factors may come in to play here. For example: What is the level of uncertainty about a particular interpretation? What kinds of imaging will help resolve it? What might help my colleague understand my interpretation and persuade them of a particular course of action? Is it worth the extra level of radiation to create a better image? Is it worth the extra injection of contrast dye? Is it worth the extra time involved? Are there other resources available to allow the clinician to act and proceed in a clinically suitable manner?, etc. These ongoing judgments about the utility of any image manipulation are bound up in the work necessary to achieve a course of surgical action.

Second, the opportunities for how medical imaging resources can be actively used in the creation of a *surgical vision* in the operating room, are bound up with the interactive properties of these technologies—from where they can be seen, from where they can be manipulated and the input mechanisms available for their control. Furthermore, there are particular features of the work and context that shape and constrain how and when, in light of these interactive properties, the imaging resources can be used during a procedure. Consider, for example, if the surgeon is holding some piece of surgical equipment or if he/she needs to be standing at the patient table—How might these affect the opportunities the clinician has for interacting with and utilizing the various imaging resources at his/her disposal.

4.2   STERILITY AND CONSTRAINTS ON IMAGING PRACTICES

One of the key factors affecting the organization of work practices in the operating room is the need to maintain strict boundaries between that which is sterile and that which is not (e.g., Katz, 1981; Johnson et al., 2011; Mentis et al., 2012; O'Hara et al., 2013). Studies of surgery by Katz (1981) in particular have highlighted the ritual nature of practices in the operating room in relation to the boundaries between different "*realms of cleanliness*"—sterile and contaminated. These ritual practices involve constrained and carefully choreographed movements among the clinicians when working within confined spaces in the room. A further key feature of the practical management of the boundaries between sterile and non-sterile is an organized distribution of labor between scrubbed and non-scrubbed personnel within different spatial zones of the room. Here scrubbed personnel touch only that which is sterile while non-scrubbed personnel touch only those things that are non sterile, coordinating their actions to interact with and manipulate the various artifacts, objects and bodies comprising the setting.

These issues are taken up more specifically in relation to the practices surrounding medical imaging within the operating room in the work of Johnson et al. (2011), Mentis et al. (2012), and O'Hara et al. (2013). The key here is that standard input mechanisms available for controlling imaging systems in theatres (mouse, keyboard, touchscreen) all require contact in interaction. As such, these input controls within the operating room are all designated as non-sterile. This presents a challenge for those surgeons and clinicians in the operating room who are scrubbed up and gloved in readiness for surgery since contact with such non-sterile devices is not possible without breaking asepsis.

In dealing with these constraints the scrubbed clinicians have a number of options available to them in order to be able to interact with and manipulate the various sources of medical imaging data within surgery. First of all, they can remove their surgical gloves and then proceed to interact with the systems using the mouse, keyboard, touchscreens, and so on. While this is sometimes used, this practice comes at a cost. In order to continue with surgery, the clinician will have to go through the process of scrubbing up again. This is a time-consuming process and to do so several times during a procedure would not be ideal with the potential for both financial and health risks to be incurred through increased procedure times. The issues here, however, are not simply to do with increasing procedure times. Rather, the time costs associated with such interactions can be a hindrance to the utilization of certain imaging resources. That is, they may be deferred until there is a natural break in the surgery, when a clinician might be happy to remove gloves anyway. Or, they may simply defer to other sources of data that may support their decisions to take a particular course of action. The time costs of this strategy mean that there is a constant need for such imaging interactions to be justified rather than more casually interleaved with other features of surgical practice.

In order to avoid the need for lengthy de-scrubbing and re-scrubbing, a well-established strategy is for the imaging practices to be collaboratively achieved by different members of the surgical team. Here, scrubbed clinicians instruct non-scrubbed personnel (e.g., radiographers and nurses) to interact with the medical imaging equipment on their behalf (e.g., Graetzel et al., 2004; Johnson et al., 2011; Mentis et al., 2012; O'Hara et al., 2013). We can see an example of this behavior in Figure 4.1. Here, a neurosurgeon and a nurse work together to navigate the image database of the PACS (Picture Archive and Communication System) display. With gloved hands the surgeon points to the area of the display without touching it while the nurse operates the mouse to carry out the actual interactions with the system.

There are times of course when such collaborative control of these imaging systems in the operating room works just fine. In particular, when the desired image manipulations are relatively discrete and articulable through verbal and gestural instruction. But such practices are also not without their problems. First of all there is the dependency on a third party. While some of the imaging systems have dedicated controllers on hand who are available to respond to the demands of the surgeon, others do not. There are significant numbers of occasions where other members of the surgical team are engaged elsewhere with other tasks and thus not immediately on hand to help. Second, the practice of issuing instructions can at times be cumbersome and time consuming. The kinds of manipulation possibilities are constrained by what the surgeon can communicate and what the third party can usefully understand in terms of the interaction request. Third, and perhaps most significant, the requirement for the clinician to issue instructions for the control of medical imaging equipment can hinder the analytic, interpretive and communicative work that can be performed through the direct manipulation of these images by the clinicians themselves.

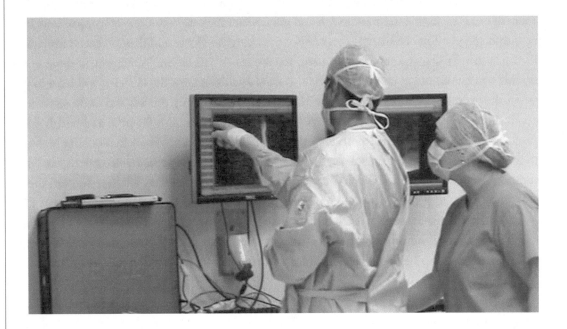

Figure 4.1: Scrubbed surgeon instructing non-scrubbed nurses to interact with the medical images equipment on their behalf.

There is evidence from the work of Johnson et al. (2011) and O'Hara et al. (2013) that clinicians, when scrubbed, do look for ways to achieve these more direct interaction possibilities with images without breaking asepsis. We see an example of this in Johnson et al.'s (2011) study of Interventional Radiologists that describes how the clinicians flick the surgical gown over their gloved hands and manipulate a mouse through their gown in order to achieve direct control over imaging resources without breaking asepsis (see Figure 4.2).

While such practices are not entirely risk free, they have become accepted practice for certain clinical procedures such as interventional radiology where the risks are small and where they are outweighed by the interpretive benefits of hands-on control of the images as well as the clinical benefits of the time saving they entail (for example, less time for the patient under anaesthetic). In other more invasive procedures, the risks of such practices are higher and therefore not readily available as an option for enabling hands-on control of the images by the clinician. In these circumstances clinicians only have the option to de-glove and rescrub available to them, which, as we have discussed, puts constraints on how and when particular imaging interactions might be undertaken.

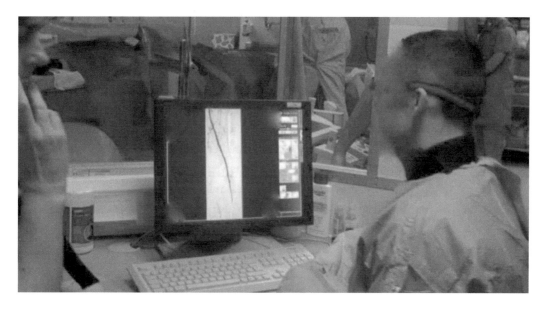

Figure 4.2: Clinician flick the surgical gown over their gloved hands to manipulate a mouse without breaking asepsis.

4.3   TRACKING THE BODY OF THE CLINICIAN FOR ENABLING TOUCHLESS INTERACTION WITH IMAGES

It is in this context of enabling clinicians to control and interact with imaging systems directly without compromising sterility that our core concerns with body tracking in the operating room come into play. In particular, it is in the ability to track the movements and actions of clinicians that opportunities have arisen to allow interaction with these imaging systems without the need for the clinician to come into contact with any form of input device—by tracking the gestures and actions of the clinician as a form of system control.

There is now a growing number of research groups and commercial enterprises that have started to explore and commercialize these opportunities for tracking the gestures of clinicians touchless gestural interaction with medical imaging systems in the operating room. The beginnings of such work came about in the middle part of the last decade with one of the earliest examples being the *contactless mouse system* developed by Gratezel et al. (2004). This system used a stereoscopic camera to track hand gestures for controlling basic mouse functionality enabling gesture control of cursor movement, pointing and clicking. While such a system served the purpose of overcoming the constraints of sterility in image interactions, it was clear such early explorations that simply emulated the mouse were limited in terms of their exploitation of the richer potential of gestural capabilities.

More sophisticated use of air-based gestures for the control of medical imaging technology began to emerge in the later *Gestix* system (Stern et al., 2008; Wachs et al., 2006). This system employed a web cam and computer-vision techniques to track a single hand of a clinician for interacting with X-ray and MRI images. The computer-vision techniques relied on pixel color matching to derive a classification of the hand position within the image frame. Instead of emulating mouse functionality, the Gestix system sought to introduce more bespoke gesture-based control for functionality such as a flick gesture for navigating through images as well as gestures for zooming and image rotation. Similar functionalities were also explored in Kipshagen et al.'s (2009) marker-free gesture tracking software for controlling the open source OsiriX medical image viewer. Stereo cameras were used to triangulate the position of a hand in 3D space. The contours of the hand were also segmented and classified into a small set of gestures for executing the navigation, rotate, zoom, and image reset functionalities of the OsiriX system.

Building on the core ideas of these initial systems, more recent years have seen a growth in the number of touchless systems being developed to support the control of medical images in surgical settings. Much of this growth in this work can be attributed to introduction of low-cost sensor and development kits such as the Kinect. The introduction of the Kinect has been significant in lowering several barriers to entry to the development of such systems including financial costs, development complexity, the need to wear trackable markers to overcome some of the inherent challenges of full depth skeleton capture from purely camera based systems. One of the first systems to take advantage of the infrared depth sensing capabilities of the Kinect sensor was the GestSure system deployed in Sunnybrook Hospital in Toronto (see Strickland et al., 2013 for an overview). Again, this system took the approach of gesture-controlled mouse emulation, which while constrained, was motivated by a number of important pragmatic considerations. In the first instance, this choice was made to enable compatibility with existing hospital PACS systems, a potentially important concern in the adoption of these technologies. Furthermore, such an approach was considered to provide value in terms of ease of use and learnability as well as reliability gains arising from the more reliably distinctive gestures possible with smaller gesture sets. This system is now being realized as a commercial concern through GestSure.

While there is elegance in the simplicity of these systems, interactions with medical images in surgical settings often extend beyond simple navigation possibilities, requiring a much richer set of image manipulation possibilities beyond the need to rotate/pan/zoom. These may include, for example, adjustment of various image parameters such as contrast, density functions (to reveal different features such as bone, tissue or blood vessels), opacity, and so on. In addition, one might look to capabilities such as marking up or annotating images or delineating points of interest for the purpose of measurement, etc. Furthermore, such manipulations may apply to whole images or more specific regions of interest defined by the clinician. With a view to providing some of these richer possibilities a number of more recent Kinect-based research has looked to developed larger gesture sets. Notable examples here include the systems developed by Gallo and colleagues (2011), Ebert and colleagues (2011, 2012), Ruppert et al. (2012), and Tan et al. (2011). Enabling these richer image manipulation possibilities is an important development but it also raises a number of interesting challenges for such systems.

One such concern is that of *expressive richness*, namely, how to map an increasingly larger set of functional possibilities coherently onto a reliably distinctive gesture vocabulary. A number of approaches have been adopted in these systems (e.g., use of modes to distinguish gestures, different input modalities such as speech and the use of composite multi-handed gestures). Using one- and two-handed tracking for example not only brings the benefits of bimanual interaction, but also enables a richer set of expressive possibilities. Both Gallo et al.'s and Ebert et al.'s systems employ both one- and two-handed gestures. Different combinations such as single hand, two hands together, two hands apart, are used to control certain image parameters that are adjusted through their respective positioning in the x, y, and z planes. More recent versions of Ebert et al.'s system have added further expressive capabilities by exploiting algorithms capable of more finger-level tracking in which spread hands are distinguishable from open-palmed hands.

Accompanying these larger gesture sets is a concern with the learnability of these systems. In an attempt to accommodate these concerns, the systems of Tan et al. (2011) and Ruppert et al. (2012) use composite bimanual gestures hands in consistent and extendable ways. In these cases, the non-dominant hand is used to denote a particular function or mode while movement of the dominant hand within the x, y, and z planes enables the continuous adjustment of image parameters. In this way common gestures can be applied across a range of different functionalities making such systems both more learnable and extendable.

4.4   CLINICAL CONSIDERATIONS IN GESTURE DESIGN

As the number of research systems in this area has grown as well the development of a number of commercial ventures (e.g., GestSure, TedCas), the concerns with these systems needs to extend beyond discussions of technical feasibility and implementation. What is additionally important with such systems is a concern with how these systems are appropriately situated within the context of the working practices or surgery and the setting of the operating room. Increasingly, we need to understand further about how the specifics of these systems—the tracking algorithms, the capabilities of the sensors, specifics of the gesture design, and other features of the interface—relate to clinical considerations of the procedures and working practices of the operating room. While a number of the systems above were developed in collaboration with clinical partners and offer a number of intriguing design choices, much of the design rationale and its relationship with clinical concerns is often not explicated in sufficient depth to allow a critical reflection on these choices. The immediate concerns of overcoming the bounds of sterility aside, it is perhaps only in the work of Strickland et al. (2013) that we begin to see some more explicit articulation of the system specification as bound to particular clinical concerns of the operating room.

In order to help us draw out some of these clinical concerns that impact on the design of these systems, we reflect on some of our own experiences in developing such a touchless interface to medical imaging systems in the OT (Johnson et al., 2011; Mentis et al., 2012; O'Hara et al., 2013, 2014a, 2014b). The focus on our own system here is for pragmatic purposes only here given that we have greater access to the particular rationale behind certain design choices. In articulating these, the aim is to raise and illustrate more general points of significance that can be applied to design and critical appraisal of these systems as they mature.

The system in question was developed to support image guided vascular surgery and more specifically complex aneurysm repair in the aorta. These procedures involve the insertion of custom-made stent grafts into the aorta, for which the clinician is guided by real time fluoroscopic imaging combined with static x-ray reference maps and overlaid pre-operative volumetric renderings of the aorta (see Figure 4.3-left).

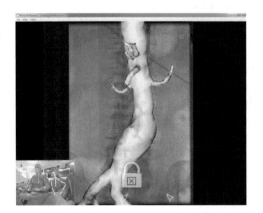 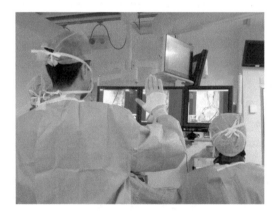

Figure 4.3: Kinect gesture system to support complex aneurysm repair in vascular surgery.

The system utilized a Kinect sensor as a touchless interface to the system to allow the clinicians to control the yellow 3D overlay shown in Figure 4.3. The gesture set allowed the clinicians to control a pointer for directing attention to particular parts of the image. In addition they could rotate, zoom and pan the overlay. Furthermore, they could place a mark on the surface of the overlaid aorta that also traces a line to the corresponding point on the underlying fluoroscopy image (see Figure 4.4-left) Finally, they could control the relative opacity of the yellow overlay to reveal the underlying fluoroscopy image and reveal the correspondence between the two imaging layers (see Figure 4.4-right).

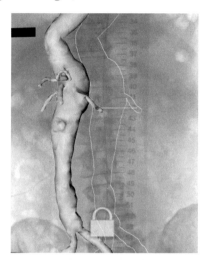 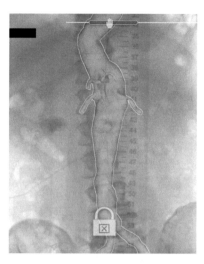

Figure 4.4: Screen shots showing (left) placed mark and ray trace; (right) opacity slider. Screen shots showing (left) placed mark and ray trace; (right) opacity slider.

In discussing further the design of the system we have oriented to a number of important clinical and socio-technical concerns.

4.4.1 CLINICAL CONSTRAINTS ON MOVEMENT IN GESTURE DESIGN

In thinking about the gesture design for our system, our concerns were not simply with creating a set of natural and usable commands. Within this context, there were a number of factors that constrain the actions of the clinician operating the system, which in turn impact on the ways that we can think about the specifics of gesture design. First of all, there is a concern with an adherence to the ritual practices of sterility that impose certain restrictions on the movements of the clinician's arms, a point that is also raised in Strickland et al. (2013). Within such ritual practices the area above and below the clinician's torso are designated as non-sterile in strict sterile practice and so surgeons must constrain their arm movements to within this sterile torso area (see Figure 4.5). This concern is a particular issue when we think about the location of the clinician. Our system was designed for use at the operating table. In the context of these particular procedures, several clinicians are working in very close proximity to each other. This imposes certain physical restrictions on arm movements either side of the controller's body as well as exacerbating the importance of adherence to the ritual practices of sterility outlined above (e.g., Johnson et al., 2011; O'Hara et al. 2013). Gestures need to be designed in order to accommodate these particular movement restrictions imposed on the clinicians. Mapping parameter control to smaller arm movements is not always a complete solution here since it increases the sensitivity of the system and amplifies any unwanted tracking errors. In order to overcome this increase in gesture sensitivity, we designed a clutching mechanism in which hands are withdrawn from the imaginary x-y plane back toward the body. When the hands are close to the body the system ignores movement and allows repositioning of the hands before moving back into the active recognition plane. This enables, for example, a continuation of the zoom gesture without the need to stretch the hands beyond the sterile region in front of the torso. The user would move their hand forward in front of their chest, then move their hands apart to begin the zoom, then move their hands back toward their chest to reposition them together to start the gesture again and continue the zoom manipulation.

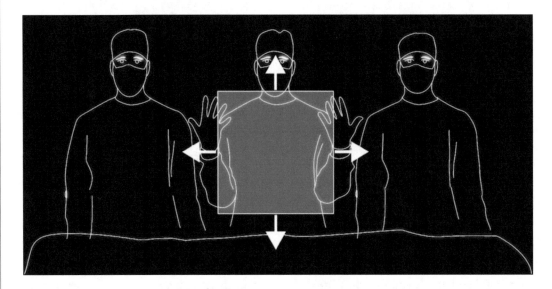

Figure 4.5: Area above and below the clinician's torso are designated as non-sterile in strict sterile practice and so gestures must be designed to work within this sterile torso area.

4.4.2   SUPPORTING COLLABORATION AND CONTROL

As well as thinking about how the design of these body-tracking systems support single-user control of these systems, it is worth considering also how some of the more collaborative features of imaging practices in these settings might also be incorporated into the design. Insights from observations in these settings would suggest a number of candidate areas for collaboration support in these systems. First, is the idea that these images are key sites for discussion and clinical interpretation by several members of the surgical team. In the context of these discussions, different members of the surgical team gesticulate and point to features within the images as they collectively work toward an interpretation. With this in mind our system provided the ability for multiple clinicians to simultaneously control an on-screen pointer with their tracked gestures (see Figure 4.3-left). This allowed them to virtually point to different features and details in the image as they collaboratively sought to resolve ambiguities and uncertainties in their interpretation. The particular significance of this arises in those situations where the clinicians are hampered in terms of their physical access to the screen. Collaboratively pointing and gesturing over an image from a distance does not enable the precision of pointing to demark the necessary features of the image under scrutiny. Getting closer to the image is not always easy and may involve awkwardly leaning over the patient. Indeed at times, the clinicians use ready to hand instruments such as catheter wires to extend their natural reach and point to details of the image from a distance (see Figure 4.6). Tracking the gestures of multiple clinicians at once provides them with remote pointing capabilities in the context of their collaborative discussion.

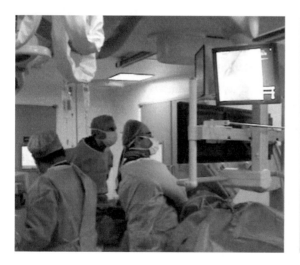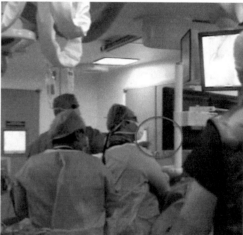

Figure 4.6: Ways of getting closer to the image for deictic reference.

Tracking multiple users in this way was of further importance in our design for the collaborative control of these images during surgery (see Figure 4.3-left). The issue here is not so much a need for multiple clinicians to be able to simultaneously control the medical images but rather to support the fluid hand over of control from one clinician to another as dictated by the in-the-moment needs of the procedure. For example, when one clinician is busy with certain aspects of the procedure, responsibility for control of the images may need to be taken up by one of the other clinicians in order to maintain the smooth flow of the procedure.

4.4.3   WHAT ACTIONS AND BODY PARTS TO TRACK FOR THE PURPOSES OF SYSTEM CONTROL

An important consideration in the design of these systems is what actions and body parts make sense to track for the purposes of developing a gesture vocabulary. For example, if we look at some of the systems we have discussed earlier, these variously make use of either one handed or two-handed gestures. Some of the decisions to use one- or two-handed gestures have been driven simply by the tracking challenges of earlier systems. But with greater sensing capabilities, there are greater opportunities for how gestural control is distributed across one or two hands. Much of the design rationale for these choices is articulated in terms of more general principles of usability that apply across any domain. For example, using one hand may be used to keep the gesture set simple and learnable while exploiting the use of two hands enables a richer set of gestures to be mapped onto a larger functional set and offers the control benefits of bimanual action during interaction.

But within the context of the surgery, there are factors over and above these generalized usability arguments that pertain to our design choices about how control functions are mapped across either one or both of the clinician's hand/arm movements. In particular here are the ways that the hands of the clinicians may otherwise be engaged when using other medical artifacts during the procedure. In vascular surgery or interventional radiology for example, these issues are recognized with the provision of foot pedals to enable clinicians to controlling initiation of x-ray imaging. This raises questions as to how many hands may be available to perform certain gestural operations at particular moments. In this respect, the design of the gestural vocabulary is not simply a question of having the right number of commands to match the functionality but also determined by the clinical context of use.

In our own system, we made use of one-handed and two-handed gestures. Working directly with the clinicians we ascertained at what points in the surgery they might require particular functionalities of the imaging interaction and whether these interactions needed to be done while holding medical instruments and catheter wires or whether they were likely to be needed at points where instruments could be put down. On the basis of this, it was decided that functionalities such as panning and zooming the image would typically be done at points when instruments and catheter wires would be put down. For other functionalities such as fading the opacity of the overlay or annotating the overlay with markers (to highlight a point of correspondence on the underlying fluoroscopy image), the clinician was more likely to need this at points where they might be holding on to the catheter, thereby only having one hand free. For these clinical reasons our system used two-handed gestures for panning and zooming, but single-handed gestures for opacity manipulation. We can see an example of how this choice plays out during an actual procedure where the system was being used (see Figure 4.7).

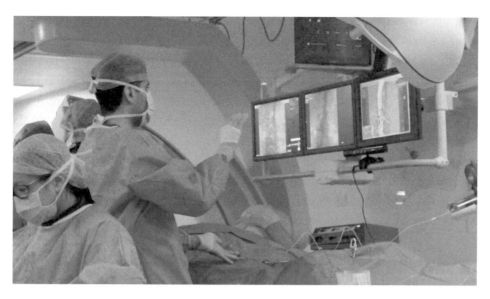

Figure 4.7: One hand to gesture and the other hand to hold the catheter.

Here the clinician is using his left hand to steady the catheter wire that extends across the length of the patient's body in order to prevent it from getting tangled. As his left hand is engaged with this, his right hand is free to perform a manipulation of the image.

The point being made here is that the mapping of tracked body movements to functional control needs to be done not just with usability concerns in mind but also with clinical significance in mind. Indeed when we consider the practical constraints of action in these settings and in particular procedures it is worth thinking further about ways in which the tracked actions of the clinicians may be combined with other modalities to achieve appropriated ways of controlling these images in ways that are both sterile and liberating to the actions of the clinicians. For example, in our system described above, we combine gesture with voice control in order to enable certain functionalities without being demanding on the actions of the clinicians. In more recent systems we have developed, we have also explored ways in which voice activation can be used as an alternative to gestural control to enable both touchless and hands-free interaction when required. Alternative choices here might be to combine gestural control with foot pedals for example. We might also consider other actions and parts of the body that might be tracked such as the head and face movement that open up opportunities for touchless body tracking as interaction at times when the hands may be occupied (c.f. Nishikawa et al., 2003).

4.4.4   ENGAGING AND DISENGAGING THE SYSTEM

An important consideration in the use of body-tracking technologies for interaction with medical imaging systems is how to distinguish between bodily action intended for gestural control and bodily action that is simply performed in the course of other movements. We have already discussed how images are a site for gesticulation in support of conversation and discussion so there are good reasons why there needs to be hand movements in front of an image that it not for the purposes of controlling interaction. Similarly, actions being performed in the context of the surgical procedure may share kinesthetic components with control gestures that consequently may unintentionally be classified as a gesture. Furthermore, in systems comprising multiple gestures in the control vocabulary, the transition from one gesture to another can involve movements in common with other in the gesture vocabulary. All of these issues raise the real possibility of the system inadvertently recognizing certain movement as system control gestures. What is key in the design of these systems, then, is the need for mechanisms to move between states of system and engagement and disengagement, reinforced with appropriate feedback to signal the system state.

The issue of correctly spotting an intended gesture in the context of other related body movements is a more general challenge for such systems irrespective of their domain of use. There are a number of ways that such challenges may be dealt with either directly or through accommodation by design. Some researchers for example have explored the possibilities of automatically determining intention to engage and disengage from the system. A good example of this can be seen in the work of Mithun et al. (2012), which looks at how intentionality behind a movement/gesture can be determined through additional contextual cues arising from the tracked body of the clinician. In this work, contextual cues, such as gaze, hand position, head orientation, and torso orientation are used to help determine whether the clinician performing the movement is actually intending to perform a system readable gesture or not. The motivating ideas behind this approach is interesting in that it extends our ideas of body tracking beyond concerns with recognizing the gestural movement itself to utilize other contextual features of the body as the context for determining gestural intentionality. While there is some promise shown in these approaches in helping avoid unintentional gestures, it remains an inherently difficult challenge to fully determine human intent on the basis of such additional contextual cues. As such, these approaches are perhaps not mature enough yet to overcome the difficulties likely to be experienced with these systems in the short term. For example, such contextual cues are likely to be similar when talking and gesticulating around the image during collaborative discussion as they are when actually intending to interact with the system.

Given this, it remains an important design concern for these systems to have explicit interaction mechanisms for engagement and disengagement from the system. The original Sunnybrook system, for example, took the approach of a deliberately unusual gesture to explicitly engage the system to recognize gestures. In their system they used above the head to engage/disengage the system with such a gesture not likely to occur in the course of other activity. The success of this kind of approach though can depend on the interactions that follow once the system engages. In the development of our system we have tried a number of different approaches with varying success.

Along the lines of the Sunnybrook example, we initially utilized a right-handed "waving" gesture to engage and disengage the system. While this in itself was unusual enough in itself to avoid inadvertent activation it did not get over the problem of *gesture transition*, whereby the movement following to prepare for the next gesture was deemed a gesture. A second approach we tried was to use a time-based lock in which holding the hands in position for a period of time. While such an approach has be successful in other domains of gestural interaction, in evaluations with the surgeons, there was a natural tendency for the surgeon to pause and inspect the image or holding a pose to point at a specific feature of the image. These naturally occurring imaging behaviors clashed with the pause-based lock gesture. In our most recent version of the system, then, engaging and disengaging control is achieved through a simple voice command which we have found complements the gesture vocabulary and works well when some discrete change of state is needed.

The ability to separate out movement for system engagement from other movements is also dependent upon the tracking capabilities of the sensor and computer vision algorithms. The ability to recognize movements and body parts at greater levels of detail open up greater possibilities for how system engagement and disengagement is designed for. More recent versions of the Kinect or the Leap motion sensor are now enabling greater fidelity in terms of hand and finger movements which extend the possibilities for how we can define the boundaries of engagement and disengagement. In later versions of our system, for example, the seemingly simple ability to distinguish between an open hand and a closed fist dramatically facilitated the ways in which system *clutching* can be implemented. When moving the arm with an open hand, the system tracks the movements but does not actually engage any of the interaction components of the system. In order to register movements or selections of the interaction components, a close fist gesture is used to more explicitly engage the system. These simple gestures to control the boundaries between engaged and disengaged movement provides the clinician with much greater freedom to move their arms and hands without consequence.

4.4.5   FEEDBACK AND MAKING ONESELF SENSED

The issue of engagement and disengagement starts to point to a further issue about how users know whether and how they are being sensed by the system. When thinking about body tracking and gestural control systems, it can be easy to idealize the ways how the gestures, as defined in an abstract gesture vocabulary are enacted and recognized in practice. While we define our gestures in terms of the bodily actions that we have to conduct, there is an important distinction to be made here between the ways the body and its actions are understood by the user and the body as sensed, interpreted and understood by the system. In the Kinect system, for example, working as a layer above the raw sensing data is an algorithmic interpretation of the body as a form of an abstracted skeleton. It is in the ways this abstracted interpretation of the skeleton enacts movements and gestures through which the control of the system is determined. In an ideal characterization of these systems there should, broadly speaking, be a 1–1 correspondence between this abstracted body of the skeleton as understood by the system and the body as experienced and understood by the controller. The reality though is that such a correspondence is contingent upon a range of factors that compromise in various ways the fidelity of this correspondence in practice.

One of the key challenges in these settings arises from the many bodies of the surgical team that occupy the room during a procedure. Given the properties of the sensor and tracking algorithm, there emerges a rather curious *zone of interference* around the controller, to the sides and behind, in which other bodies in the room come to impact on the system performance. We can see in Figure 4.8, a view from the Kinect sensor of the clinicians working at the patient table. Key here is the clinicians are working in close proximity and wearing similar color clothing. The clinician with the raised hands is attempting to control the system and, from his perspective, correctly performing the gesture in question. However, the system here is producing an abstracted skeleton that extends across the arms of the colleague to the side of the clinician in control. What the system sees as the controller's body in this situation, then is different from the controller's actual body. As a consequence, the enactment of gesture as understood by the clinician does not correspond with the enactment of gesture as understood by the system.

Figure 4.8: Abstracted skeleton of Kinect sensor extends across clinical colleagues working in close proximity.

Other challenges to the sensor and computer vision algorithms may also arise from colleagues in the background. As they move about in the performance of their work, they come in and out of the sensor frame with the potential to interfere with the algorithmic determination of controller's body. In part, we can characterize these issues as an interesting set of challenges to be solved with evermore sophisticated tracking and machine learning techniques. Or we can think of other types of sensor technologies where the impact of bodily proximity of colleagues may be less apparent.

But we can also look to ways at how we accommodate and account some of these existing zones of interference in other ways through appropriate interaction design interventions. One such intervention that can be considered here is the provision of some form of feedback regarding what the computer vision system is seeing. An example of this can be seen in Figure 4.3-left. In the lower left-hand corner of the screen is a real time picture-in-picture view of what is being seen and interpreted by the computer vision system. By providing this view, the clinicians can determine whether their bodily movements are being correctly interpreted by the system or whether there are problems of the type illustrated in Figure 4.8. Furthermore, such feedback provides insights into exactly how the bodily position and movements of those around the controller are impacting on the system. This enables the clinicians to reconfigure their position and movements in order to avoid interfering with the controlling clinician (see O'Hara et al., 2014a).

4.4.6   COARSE VS. FINE-GRAINED CONTROL

An important consideration for the use of body tracking and gestural control within these settings concerns the control sensitivity required for different kinds of functional interactions. That is precisely how the movements of the body need to be tracked in order to achieve appropriate levels of control. Observations in the operating room point to different levels of image manipulation some of which are amenable to large-scale and coarse-grained tracking and others for which more fine grained control is necessary (e.g., Johnson et al., 2011). For instance, selecting the next image in the sequence is something that is relatively amenable to rather coarse grained-tracking of bodily action of the clinician. But other tasks require greater levels of control precision. In interventional radiology, for example, the clinicians will select specific parts of an x-ray image in order to selectively apply transformations to these parts of the image. Similarly, specifying points and regions in an image for the purposes of clinical measurement also require more fine grained levels of precision.

Such fine-grained activities can be difficult in light of lower levels of accuracy in hand tracking inaccuracies owing to the inherent problems of jitter in distal pointing and selection tasks. Such issues of jitter are a feature of distal pointing more generally and attempts to accommodate these include tracking filtering algorithms, zoomable locations on the display and ability to control cursor speed (Kopper et al., 2010; Vogel and Balakrishnan, 2005; Frees et al. 2007) which are all essentially different ways of altering the *control-display (CD) ratio* (correlation between movements of the physical input device and the on-screen cursor). What is key to our concerns here though is that coarse and fine grained body-tracking tasks require different approaches to the filtering algorithms in order to create the experience of smooth and precise pointing. That is the bodily movements and actions need to be tracked and represented differently for these different kinds of manipulation tasks. For some interactions it is fine to remove or filter out features of the action in order to achieve the desired interaction experience. But for others removal of these features would inhibit more precise levels of control. What it means to track our bodily actions and gestures then can mean very different things according to the circumstances.

Various tracking filters and algorithms attempt to compensate for these different tracking requirements by adaptively manipulating the CD ratio for controlling the accuracy-lag tradeoff. Most notable here are the PRISM (Precise and Rapid Interaction through Scaled Manipulation) (Gallo et al., 2011) and Adaptive Pointing (König et al., 2009) techniques. The PRISM algorithm is based on the assumption that if user is moving slowly he is likely to have a precise target location in mind. The speed factor is computed in a 500 ms window, which while limiting jitter, can also introduce lag. Furthermore, movement is independently scaled along the X and Y axes, providing additional jitter resistance when moving on straight line at the cost of an increased difficulty when moving diagonally. There is also an offset between user movements and cursor movements that is disconcerting while sudden modulations of the CD ratio can lead to unnatural cursor speed increases. The Adaptive Pointing algorithm (König et al., 2009) attempts to overcome the problems by adjusting the CD ratio based on the pointing device speed and the accumulated offset which helps avoid abrupt switches between precise and absolute movements. Independent axis scaling is also employed to provide more tolerance to noise when moving on straight trajectories. Due to the different jitter reducing mechanics, it is more difficult for the user to keep the cursor steady compared to PRISM. Moreover, if the pointing device moves slowly and the maximum offset value is reached, the algorithm tries to recover it even though the user is still moving slowly. The e-PRISM algorithm attempts to combine the positive aspects of PRISM and Adaptive Pointing in order to create a filter that allows accurate pointing at a distance without sacrificing the feeling of operating an absolute device. The e-PRISM adopts a velocity-based transfer function to allow for accurate users' cursor movements on the screen when precise pointing is required. Therefore, when the user is slowly moving his hand, suggesting that he has a precise cursor location in mind, the speed of the interaction is scaled down. As the hand moves faster, the transfer function changes to provide an absolute, unconstrained cursor movement on the screen. The transition between precision pointing mode and unconstrained absolute pointing is seamlessly and smoothly triggered by changes in hand velocity. In the precision mode each movement is scaled independently over the X and Y axis, working as an axis locking mechanism. This mechanism provides better jitter removal when moving over a straight line. E-PRISM, as the Adaptive Pointing technique, aims at not to undermine the user's perception of absolute device operation. This is achieved using the recovery offset as a factor when switching between precision and unconstrained pointing modes. Furthermore, when operational mode completely switches to unconstrained pointing mode, and precision is no longer required, the pointing device position is directly projected on the display surfaces thus reinforcing the user perception of absolute device operation and resolving the shortcomings of both Adaptive Pointing and PRISM.

The key point of discussing these algorithms is essentially to highlight the different ways in which the tracked actions of the body are deliberately included or ignored in any computational representation of the sensed data. The underlying algorithms here are selective in their use of data accommodating particular characteristics of the sensing and tracking capabilities to achieve particular interactive effects These selective choices about the which data to keep and which to ignore, in turn are contingent upon the application and context of use. In this respect, the way we understand the body and its movements here is not an absolute or singular thing but rather something that is bound up to the specific control requirements of the scenario in question. Indeed, this is a theme that we also discuss in relation to other domains in which such body-tracking technology may be deployed.

4.5   BODY TRACKING, GESTURE, AND ROBOTICS

Alongside developments in medical imaging technologies, surgery is also being massively transformed through the introduction of medical robots. In particular, we have seen the rise of robot assisted minimally invasive surgery with key technologies such as the DaVinci machine. As well as these, we are seeing research efforts exploring other aspects of assistance with the operating room such as the use of intelligent robotic scrub nurses (e.g., Kochan, 2005; Carpintero et al., 2010; Wachs et al., 2012; Jacob et al., 2011, 2012). For interacting with such systems, various forms of body tracking and gestural control have been considered. The motivations for these relate in part to the arguments presented above about sterility but there are further considerations that relate to appropriate forms of multimodal control within the context of particular surgical procedures.

In the Scrub-Nurse Robot (SNR) project (Ohnuma et al., 2006; Miyawaki et al., 2005; Nomm et al., 2007, 2008), the motivations for developing a robotic scrub nurse support relate to compensating for a shortage of experienced scrub nurses within the operating room. In developing the robot there is recognition of the experience and knowledge of the scrub nurse in the process and timing of instrument transfer to and from the surgeon. Within this context then, the motion analysis is intended to support the coordinated activity of the robot in the surgeon in the transfer of medical instrumentation between them. To enable this coordinated activity between the robotic nurse and the surgeon, the motion analysis components then are not about the giving of commands but rather about determining particular activities of the surgeon. Based on the observations of surgeon's motions and actions that a trained scrub nurse, the system looks to segment the surgeon's motions into a number of discrete types. This segmentation is based both on the actions as meaningful to a human as well as the ways that different motions can be meaningfully distinguished in mathematical and statistical terms. The key motions in this segmentation are:

- *inserting*—motion sequence from when the instrument is received by the hand of the surgeon to the point where the surgeon inserts the instrument into the patient;

- *working*—motion while the surgeon is using a surgical instrument in the course of the procedure;

- *returning*—motion sequence from when a surgical instrument is removed form the patient to the point it is handed to the scrub nurse or returned to the instrument tray;

- *waiting*—motion sequence between the surgeon's release of one instrument to the point where they receive the next from the scrub nurse; and

- *get*—the specific motion when the surgeon takes the instrument form the nurse.

Characterizing these motions was based upon the sequential x, y, and z positional coordinates of the surgeon's wrist, chest, and elbows. The system in this instance relied upon the use of dedicated colored markers on these part of the surgeon's body although one can might also imagine how more contemporary techniques and depth sensing capabilities offer markerless ways to capture these specific movement coordinates. The significant illustrative feature of this work is in its attempts to recognize and interpret the existing repertoire of movements and using this as a means to determine a coordinated system response. This starts to point to an interesting set of possibilities about how such body tracking and motion recognition of the surgeon might be used to anticipate intent and reactively configure features of the surgical environment in more optimal ways (e.g., Wachs, 2009).

Other efforts to develop robotic scrub nurse assistants have adopted a different approach to issue of human-robot interaction in the surgical room. In developing the GestoNurse (see Figure 4.9) system (Jacob et al., 2011, 2012; Wachs et al., 2012), the motivations of automating the transfer of instruments is to allow the surgical team greater opportunity to focus on other complicated tasks such as maintaining the sterile environment, preparing surgical supplies and monitoring the patient as well as attempt to reduce errors and delays arising from possible communication failures between the operating staff. The concerns of the researchers with this system is not so much the automatic recognition of the surgeon's actions but rather with offering an appropriate method of instructing the system to pass particular instruments. While some other attempts in this space had used voice recognition for this purpose (e.g., Kochan, 2005; Carpintero, et al. 2010), the developers of the GestoNurse system looked to overcome some of the challenges of voice recognition in the OT by resorting to hand gesture recognition with a Kinect depth sensor. The system identified hand and fingertips with a view to allowing the surgeon to perform static hand-based gestures that could be recognized with greater speed and accuracy than comparable voice based systems—another potential advantage of this form of body tracking for human robot interaction.

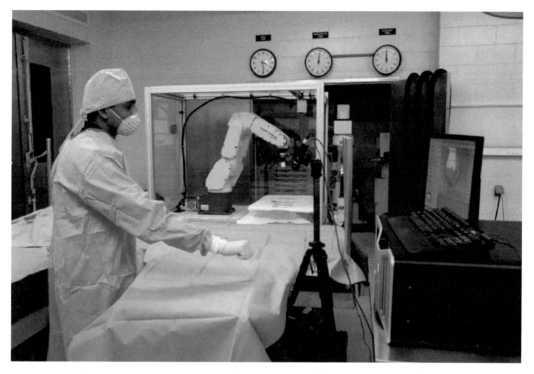

Figure 4.9: GestoNurse.

4.6   INCREASING INTERACTION BANDWIDTH THROUGH INPUT MODALITY

A further area of robotic assistance within surgery can be found in endoscopic and laparoscopic procedures. These procedures involve the insertion of a scope either through existing orifices in the patient's body or through dedicated incisions. The scope provides a real-time camera view inside the body. Key in these procedures is the need to control the position and orientation of the camera. Traditionally, the control of these systems was undertaken by a dedicated assistant under the instruction of the operating surgeon allowing the surgeon to focus on other actions and instrument manipulation in the context of the procedure. As in our earlier discussions of imaging control, the communication of instructions can come with certain challenges and can introduce extra possibilities for error. In addition, the views produced by the assistant may not always be optimal with further issue of vibrations arising from the natural tremors of the assistant's hands. The introduction of robotic laparoscopic positioning systems, such as AESOP and more recently the Da Vinci system overcomes some of these difficulties but some difficulties still remain. In particular, there is an additional burden of control with these systems. So while there are input options available for the surgeon to control these systems, such as foot pedals or hand-controlled input devices, these input modalities are already being used in a variety of other surgical instruments. As such their operation requires the removal of hands from the main instruments and shifts of attention from the display to these other controllers (e.g., Nishikawa et al., 2003; Staub et al. 2011). Overcoming this control burden and opening up new input modalities is then a central concern for the operation of these systems in order to allow them to be used effectively alongside other systems in the operating room without the need for constant transfer of attention and body parts from one point to another.

Alongside explorations in voice control (e.g., Allaf et al., 1998; Vara-Thorbeck et al. 2001) a number of researchers have also considered various forms of body-tracking solutions. In particular, there has been a focus on the head and face in terms of its position and orientation as well as more specific features such as eyes and mouth. Perhaps one of the most well-known examples here is FAceMOUSe (Nishikawa et al., 2003). Unlike earlier systems, which used head movements for initiating discrete commands (e.g., Finlay and Ornstein, 1995; Finlay, 2001) and required complimenting with use of a foot pedal, FAceMOUSe was designed to enable the control of both discrete *trigger* commands and continuous control of the laparoscope and without the need for a foot pedal. In FAceMOUSe, face yaw, orientation, and tilt are used to provide the input degrees of freedom to control mode changes, pan and tilt and camera insertion/retraction.

Again, there are certain pragmatics of tracking that come into consideration here. In particular there is the issue of the surgical paraphernalia that covers the surgeons face such as the surgical mask. In the FAceMOUSe system, then, the use of a standard camera meant that an iris tracker combined with the tracking of a rectangular marker placed on the surgeons cap were the basis of determining the orientations and motions of the head necessary to achieve the control. More recent efforts in this space have been able to achieve the same levels of control in more sterile marker free ways. In Wachs (2009), for example, the depth-sensing capabilities of a Kinect camera now mean that these same degrees of freedom of head control can be used without markers to achieve hands-free and foot free control of an endoscope.

The motivation of this work to increase the surgeon's interaction control bandwidth remains a good one and an important concern for how body-tracking capabilities are considered in the creation of new interaction possibilities for surgeons. But it is also one that must continually adapt to an evolving set of restrictions on the surgeon's bodily capabilities and new procedural practices. If we consider one of the most successfully adopted technologies within this space, namely the daVinci system, its use imposes certain pose restrictions on head movements. As such, the kinds of motions utilized in the above head tracking systems may not always be appropriate as control modalities for hands free and foot free control within these systems. In these contexts, other features of the head or body may need to be considered as tracked possibilities in the production of greater interaction control bandwidth (e.g., Gomez et al., 2009).

4.7   CONCLUSIONS

In this chapter, we adopted a somewhat different perspective on the role of body-tracking technologies in healthcare by making the clinician the primary subject of the sensing technology. A key motivation behind this kind of use in healthcare has been the challenge of interacting with ever more complex equipment while maintaining sterility. The use of gestures in this context then are not so much concerned with simplifying interaction or making it feel more natural to the clinician as is often cited in relation to gesture input but rather with significance of controlling medical equipment without touching (O'Hara et al., 2013). The important point to stress here is that the aim of these touchless systems is not to make the environments of their use more sterile. Indeed, clinical staff already orient to sterility in very strict terms but in doing so their imaging practices have to be organized and configured in particular ways, such as by clinicians instructing third parties to manipulate images on their behalf; by degloving and rescrubbing; or by simply restricting and avoiding certain types of image interactions during the surgery. By avoiding these cumbersome routines and workarounds, touchless interaction for the clinicians in this context enables them to reconfigure the socially constructed practices of seeing or *professional vision* (Goodwin, 2000).

Of significance in this context is that the touchless interaction alters the *cost structure* of information access. For a clinician, the decisions to access data and images is not prescribed but rather are an in the moment judgments. Such judgments are bound up both in what the data might enable clinically as well as what it entails to enact it, e.g., time, effort, downing tools, exposure to radiation, exposure to contrast dye etc. The gesture technologies here reduce some of the time and effort costs associated with information access under conditions of asepsis. In doing so, they are not just providing faster ways of doing the same information access behaviors. Rather, by lowering the cost structure of image manipulation, it offers new opportunities for how image manipulation will be incorporated into surgical practices. Most notable here perhaps is a reduced dependence upon indirect image manipulation through the instruction of third-party colleagues. Enabling closer direct control over the timing of image manipulations by the clinician affords richer forms of image understanding allowing them to be better combined with other information resources in the production of medical interpretation. Of significance is that the close timing of these manipulations made possible by direct clinician control allows them to be aligned with shifts of attention across both different layers of the imaging system and across the other displays.

The incorporation into practice is a critical perspective on how these body-tracking technologies may play out in surgical practice. It is also fundamental to how we consider them in design. From a practical perspective, we can see how this has a bearing on things such as gesture design for these technologies. For example, contextual restrictions on particular movements are apparent here such as the need to move arms within the sterile region of the torso, or the limited space afforded by working in close proximity to clinical colleagues. In addition, the use of one-handed gestures acknowledges a value in interaction while engaged with other surgical paraphernalia. Above and beyond, this are additional complexities arising from the broader relationship of the system to the context of other artifacts. For example, there are constraints on position and bodily orientation of the clinician arising from their orientation to the patient, other colleagues, and other equipment—one cannot always assume an ideal orientation in relation to the sensors. In the case of a depth-sensing Kinect, for example, these other contextual factors impacting on the embodied conduct of the clinician have a huge impact on where interaction via body tracking can take place. With this in mind, some of the different spatial characteristics of other tracking technologies (e.g., inertial sensors worn on the surgeons body) may afford new kinds of opportunities here. Interestingly, this attention to spatial concerns in relation to body-tacking sensors does not just highlight constraints on use in context. Indeed, it also opens up some new opportunities that relate to the ability to manipulate images from a distance—allowing us to rethink new architectural arrangements for the operating room.

By considering these same technologies in the context of radically different settings and user populations, key points of contrast and similarity can be revealed. With the emphasis of technology in these settings on tracking for interaction rather than tracking for measurement, there are different assumptions that pertain to underlying computer-vision algorithms. So, for example, in this context, we see the application of particular data filters that make sense from an interaction point of view but not from a measurement point of view. In one situation, one wants to keep certain data points, while in another situation, with the same technology, one looks to throw those data points away. This is a critical insight about the nature of these technologies. It is all too easy to be seduced by a fixed objectification of the body that seems available with these technologies. However, it is important to recognize the contingent forms of the body represented as data and the assumptions bound up in these representational choices.

What we see again in these examples is that these technologies ultimately become situated in particular forms of social and collaborative practice. In the settings of the operating room, the surgeon is not a single body acting in isolation but is deeply engaged in collaborative activity with other clinical colleagues. The implications of these colleagues is important to consider here in terms of how they relate to particular characteristics of the body-tracking technology. In the first instance, we see how the presence of others can create all sorts of constraints and difficulties with the *in situ* use of these systems. For example, colleagues in the background or those working in close proximity can all have an influence on the sensing performance of the system. In the work presented here, we see how it is possible to attend to these issues in design—for example, in the form of providing feedback to the clinicians about what the camera sees and what the system interprets as a tracked body. This points to a broader concern in our thinking about body-tracking technology that relates to how we make interpretable what is being seen by the system. By making these things visible and understandable to those using it, we enable collaborative colleagues to appropriately configure themselves in relation to each other and the system to achieve meaningful outcomes in practice.

The collaborative rather than individual perspective is also significant in how we understand some of the key value points of these systems and consequently what we emphasize in design. The same direct control over timing alluded to earlier extends also to the ways that these image manipulations can be incorporated into and aligned with collaborative discussion among colleagues. The timing of the manipulations allows them to be interleaved and combined with features of talk in ways that draws attention to particular features of the image. These direct manipulations by the clinicians is what gives the images meaning in context.

While our examples here have certain specific features in terms of the settings, artifacts, imaging systems, and collaborative practices, the aim is to illustrate a set of more general considerations for the design and evaluation of these systems. In particular, our attention was turned further to the interactive properties of these body-tracking systems and how brings a new set of design implications and values. Moreover, our intention was to extend our thinking beyond concerns with sterility and work efficiencies and look to new opportunities for how they become constitutive of the social practices of surgery.

CHAPTER 5

Conclusions

5.1 INTRODUCTION

This book has highlighted the exciting possibilities that sensor technologies are opening up in the healthcare space. From the assessment and monitoring of medical conditions, to new opportunities for rehabilitation, to innovations in interaction in the operating room, body-tracking technology makes possible a whole new world of applications and systems. One important aspect of this change is that advances in the availability, diversity, cost, robustness, weight, accuracy, and reliability of these sensing systems have meant that we are now seeing a move out of controlled, specialist laboratory settings into real-world contexts, with all the messiness and complexity that entails. Whether it be the clinic, the operating room, the home, or out and about in the world at large, developing healthcare applications for everyday life presents a new set of challenges for researchers, designers and developers of these systems. It is our hope that the research we have discussed and the in-depth case studies we have presented help to give a flavor not just of the nature of those challenges, but also how we can begin to address them.

A key message from this body of work is that the design and development of these applications and systems must be based on multidisciplinary efforts that pay attention to the relationship between technical, social, design, and clinical concerns. As we have shown, the issues are rich and complex, and many different factors and stakeholders must be taken into account. In this respect, developing body-tracking applications for healthcare is no different from developing complex systems for other real-world domains. Indeed, the approaches and methodologies described in this book reflect many techniques which have been developed under the umbrella of Human-Computer Interaction: participatory design, prototyping, iterative design, field deployment, and laboratory studies are common ways of putting users and the context of use at the center of our endeavours.

However, we would also argue that using body-tracking technology comes with its own special challenges. As Bellotti et al. (2002) highlighted some years ago, new kinds of sensing systems should require designers and developers to ask new kinds of questions about the systems they design. As we move away from the realm of graphical user interfaces and screen-based interaction, the "interface" may be more difficult for the user to ascertain and understand. Whether we track bodies though wearable sensors, computer-vision systems, or by incorporating other modalities into the interaction such as voice, it is less clear to users where the sensors are, what kinds of data they capture, what actions are possible, how the data are analyzed, and what the mapping is between a user's actions and the system response. There are many reasons for this, including the fact that sensors may be invisible to the user, that the algorithms processing the data may be hugely complex and based on implicit assumptions, and that there are few conventions for interaction designers or indeed users to fall back on. As our case studies have shown, it is crucial in developing body-tracking applications that patients, health professionals, surgical teams, and caregivers need to understand how and when they are being sensed, how to issue commands, how to configure their actions to recover from mistakes, and so on if they are to effectively incorporate these new systems into their lives.

5.2   CONTEXTUAL DESIGN

How then do we go about the task of developing a body-tracking application or system? The argument in this book is that the process requires a deep understanding of the intended context of use of the system, as well as the users (patients, carers, health professionals, teams) the system is targeting.

At a simple level, it is important to establish from the outset *where a system is intended to be used and by whom.* Is it for use in a clinic by health professionals and their patients, at home by people self-managing their condition, or by specialist surgical teams in an operating room? These are vastly different situations, calling for very different kinds of design decisions. In the case of assessing multiple sclerosis in a clinical situation, the system needs to be designed such that it does not undermine the delicate interaction between clinician and patient; in the case of home use, the system needs to motivate people to engage with it in an ongoing way and to accommodate the everyday routines and demands of the extended user network in the home; in the case of the operating room, the system needs to acknowledge the collaborative context of surgical teamwork as well as the embodied nature interactions with the medical images that constitutes the practices of their professional vision

Another important set of factors has to do with the *clinical goals of the system.* This book presents case studies that illustrate the range and diversity of these goals. We have seen that in the assessment and diagnosis of multiple sclerosis, the output of the system should augment professionals' ability to make judgments about disease progression and treatment, rather than aim to remove them from the process. This calls for a system that produces meaningful output that health professionals understand and trust. When it comes to rehabilitation, a goal is to train people in new skills that can eventually be transferred to real world activities. However, there are social and emotional issues at play that are crucial to successful treatment: in the case of chronic pain, for example, the system needs to help patients be more body-aware and gain confidence in their movements. Likewise, for the elderly, bolstering confidence is key, but teaching self-management is important for ongoing use of the system. In this case systems based on games give users a sense of achievement and even social engagement, both of which motivate their use in a home or a care home context. In the operating room, there are goals that have to do with balancing the need for maintaining sterility with the need for surgical staff to feel more "in control" of the data they need to carry out procedures, and in such a way that supports their professional vision.

Decisions about the intended context of use and clinical goals are ideally decided upfront as changes can have many knock-on consequences for design. This may be challenging in itself in multidisciplinary teams. But even once there is agreement, we have seen that there are many choices and design decisions which follow about different aspects of the system, ranging from the technology used, through to determining new protocols and practices that need to be developed to support the adoption of the new technology. Let us summarize.

5.2.1 SENSOR TECHNOLOGY

As we have outlined in this book, the range and diversity of sensing technology is growing, with vast improvements in resolution, accuracy, weight, power, robustness, and reliability. As such, we have only been able to give a snapshot of the current state of the art, and sketch out general trends in what constitutes a fast moving field. What is important to understand, however, are the ways in which the technological landscape provides both new possibilities as well as constraints on the kinds of systems that can be built.

To some extent this is about the performance characteristics of different sensor technologies. What exactly will they capture: the geometry of movement, force, and pressure? And, with what accuracy, precision, resolution, and reliability? Sometimes it makes sense to capture data from a variety of different sensors, as we saw in the Emo and Pain study in Chapter 3 where 3D motion capture, EMG, and facial expressions were used. Other times, more is not necessarily better, as in the balance study reported in Chapter 3, where the center of pressure on a balance board provides a much better means of both measuring and controlling balance in a game than full body tracking. So choice of sensor must be appropriate for the aims of the system.

More than performance characteristics, sensors also have other affordances. Camera-based systems are often seen as unobtrusive, since they are marker-free and require no special wearable equipment. However, they require a relatively clutter-free, dedicated space in order to deliver noise-free data and so as not to undermine interaction with the system. As such, they are better for interactional spaces which are spatially quite well defined. By contrast, activities in everyday life are often not like this, so wearable sensors may well make more sense for these kinds of applications, capturing body movement data across a range of different environments and activities.

5.2.2 DATA AND ALGORITHMS

Whatever choices are made with regard to sensor technology, it is important to remember that, while there is a general trend toward more accurate, higher fidelity data, movement capture will always be selective. Moreover, deciding what should be captured and the algorithms we develop to process and analyze that data have central impact on what these systems are capable of, as well as how they shape the resulting system response.

Data can serve as a direct measure of a movement (such as sensor data about position, velocity, or acceleration), or can be used to make inferences about symptoms or movements of interest. In the latter case, machine-learning techniques are often employed. Direct measures are often easier to understand, such as measuring the amplitude of tremor for a particular movement. By contrast, more complex algorithms, such as those developed through machine learning, can help to integrate sensor data, and even discover features of the data that may be important, but at the same time they can be more difficult to understand. Further, we have seen that it is often important for users of these systems to understand how the input is analyzed. As we saw in the case of Assess MS, for example, mapping the output of the system onto clinical judgments about patients is critical in order for them to trust these systems. It is not enough to develop algorithmic "black boxes," no matter how clever they seem to be.

Another challenge with developing complex algorithms is that they may be too specialized to use in different contexts. For example, the software development kit shipped with the Kinect depth camera was originally designed to work in a home gaming context, and so was designed to capture whole body movement of individuals rather than members of a tightly packed surgical team with mainly only the team's torsos visible above the operating table (Chapter 4). The algorithms for the operating room had to be re-written. Further, gaming algorithms give rise to a fair amount of jitter in the segmented, skeleton representations that the algorithms produce. This of course is not ideal for trying to isolate tremor arising from MS patients' movements as against noise in the data. In Chapter 2, we described how we needed to develop new algorithms in order to isolate the important aspects of the signal from the noise.

The bottom line is that while capturing more and better "quality" data might be seen to be ideal, in designing for real-world healthcare it is more important that data capture and analysis is both clinically relevant and appropriate for the interactional context. As we outlined in Chapter 1, a data set does not provide some self-contained representation of the moving body or some kind of objective reality, but rather provides a resource for interpretation. Algorithms help shape that interpretation, but even then there is a great deal of choice about how one represents the output of these algorithms back to clinicians and patients. If we consider the case of Assess MS, it may be more important to be meaningful than to accurately reflect what the algorithms actually do.

5.2.3 DESIGNING MOVEMENTS

We have also discussed in depth how important it is, up front, to select and plan the kinds of body movements that need to be captured and assessed. This is crucial in order to optimize what is captured, how the data are processed and represented, and how the system responds. Why do movements have to be designed? Ad hoc movements can of course be captured in the course of everyday life using sensor technology, but it is difficult to control for contextual factors and the messiness of real life. This approach has to be linked to a deep understanding of real life activities to show where clinically relevant movements are likely to occur and to describe the form they may take.

In practice, most health professionals already use designed movements that are clinically meaningful for their area of expertise. This can range from finger tapping to test for recovery from stroke, capturing incidents of freezing gait for Parkinson's disease, to sit-to-stand movements for people with chronic back pain. For some conditions, there are whole batteries of tests designed to assess different aspects of motor ability as well as cognitive and emotional factors. A good example of this is described in Chapter 2, in the context of multiple sclerosis. Additionally, as we saw in Chapter 3, balance is not purely a motor task but interacts with cognitive tasks such as problem solving or attention switching. Hence, the games designed for balance training reflect this. Another important factor is the relevance of these tests for daily life. Some tests, for example, for MS patients, aim to assess activities more common in daily living, such as drinking from a cup. For chronic pain sufferers, the fundamental act of reaching forward is seen as important because it maps onto many daily activities.

But we have also seen that even in more controlled environments, and in the context of movements which are standardized and well understood by health professionals, movements often need to be carefully selected or modified in order to optimize for the sensors and algorithms used, and to be easily carried out by users working with the sensor system in context.

The kind of sensing system can limit the movements that are possible. For example, depth cameras are more sensitive to lateral movement across the field of view than movement away from and toward the cameras. This creates problems for measuring movements such as gait where walking usually occurs toward and away from a camera to stay within the field of view. Likewise, the restricted field of view means that movements must also be constrained if they are to be seen by the camera (as we saw both in the case of clinical assessment and surgical teams). Seated movements are therefore especially good in this regard.

More than this, as we saw in Chapter 2, even in clinics where movements are performed according to well-understand clinical protocols, there is significant variability in how they are performed. This creates problems for algorithms: for example, in the Assess MS project, clinicians are more forgiving of variability in movement tests than machine algorithms. Without standardizing the movements, it is a challenge to capture training data such that the algorithms can distinguish between normal performance variability and variability that is symptomatic of the underlying disease.

The underlying data processing aside, another set of constraints has to do with designing movements for the interactional context. In Chapter 4 we outlined various constraints on the design of the gesture set for interaction in an operating room. This included the following.

- The use of one-handed interaction in order to accommodate for the fact that the other hand may be holding an instrument. Two-handed gestures are usually more effortful and thus were better for less frequent commands.

- Movements needed to be restricted since the sterile region within which the surgeons could move was also restricted. In the surgical system we describe, this involved implementing a clutching mechanism so that the hand could be brought back to the body without engaging the system.

- Designing movements to be ergonomic in that they did not require sustained or precise in-air movements.

- The gesture set also had to be easily learnable. We found that gestures that were continuous and spatially analogous to what one would do in the physical world worked well here, such as the gestures for rotating, zooming, and panning.

In general, movement design will need to be based on a careful analysis of ongoing activities, taking into account the clinical goals of the system.

5.2.4   INTERFACE AND INTERACTION DESIGN

All of the above considerations—the users, context of use, clinical goals, underlying technology, and movement set—feed into the design of the overall interface for the system. Here is where iterative design processes are especially important. From testing early concepts (perhaps in sketch form) with users, through to low-fidelity prototypes in the lab, to high-fidelity, more robust prototypes in real settings, these kinds of user-centered design techniques are well known within the field of Human-Computer Interaction. Design of the interface which in turn shapes the unfolding of interaction needs to consider everything from what kinds of input that are appropriate for the system, through to how users manage or engage with the system in an ongoing way, to the kinds of output or system responses are best for the particular situation.

We have already talked about the importance of carefully selecting the right sensor technology for key movements, but other kinds of input will likely be required by the system, whether this is aimed at patients, clinicians, or caregivers. In the case of the system for vascular surgeons (Chapter 4), we found that supplementing gesture with voice, especially in the case of discrete commands, made interaction more efficient and less tiring. Voice was used for changing modes, for locking the screen and so on, whereas as gestural controls worked better when images needed to be manipulated in spatial, continuous ways. In Assess MS (Chapter 2), both touchscreen interaction as well as a remote control gave clinicians freedom and flexibility in input supporting either interaction up close to the screen or a distance when they needed to be closer to the patient in order to physically support them. Voice input would have been another option here for remote interaction. What we find is that no one kind of input is best: choice of input, whether it is for commanding a system, navigating through data, or inputting data, must take careful account of the affordances of that input modality. Crucially, one must also consider how different input modalities might work together.

There are many different aspects of designing ongoing interaction with a system. In Chapter 2, we described how it was important to design the interface such that clinicians could fluidly move through different assessment tests rather than being locked into a prescriptive workflow. Balancing the right amount of structure with the ability to skip over or repeat movement tests took some time and iterative testing to get right. In Chapter 3, engagement in the case of balance training drew heavily on exploiting what game designers implicitly know. This includes the importance of personalizing the interface according to ability, giving good feedback on performance, providing a fun and immersive interface, and providing a reward structure to reinforce achievement. All of these factors help keep users engaged. In turn, this rehabilitative system had all kinds of benefits beyond the physical, giving its elderly users a sense of achievement as well as fostering social interaction. Finally, in developing the system for vascular surgery described in Chapter 4, it was important to design the system such that gestures in the normal course of teamwork did not unintentionally engage the system. This was dealt with by ensuring that surgeons needed to enter a "control" mode that was reinforced through the use of a colored border on the display screen with accompanying icons to confirm how the system was interpreting the gestures. Voice control to move in and out of control mode turned out to be a good way of quickly changing modes.

More generally, feedback to users of body-tracking systems is vitally important for other reasons. Good feedback, whether visual or auditory, helps users know whether they are being sensed, and how captured data are being interpreted by the system. In the case of chronic pain, the use of audio proved an effective way of reinforcing patients' awareness of their own movements and letting them know when they had achieved goals. Sound output had the added benefit of not requiring fixation on a display, thus allowing freer movement with the added benefit of facilitating mindfulness and relaxation. Good feedback can also help users capture optimize use of these systems. In the Assess MS system, for example, a view showing what the depth camera was "seeing" helped clinicians clear clutter in the environment and position patients correctly in order to provide better quality data for the machine learning algorithms. In the surgical setting, a view of what the camera was seeing helped the team diagnose problems when the system stopped responding as expected. In particular, a view showing how the machine learning was segmenting the skeleton representations of the people in view also highlighted when those interpretations were wrong, which in turn supported the surgical team's repositioning themselves to help the system work better.

Finally, the representation of the data can be important for assessment and diagnosis of clinical conditions. We saw in the case of Assess MS that there are many ways one can represent the output of the machine learning algorithms. In order to be meaningful, these representations need to have some connection to how a clinician assesses a patient. This might mean that it need not be a true reflection of the analysis, but rather shows the link to a particular patient's movements and draws attention to aspects of the movement that affect the output. Of course, in addition to this, system output can also help clinicians see patients in entirely new ways, such as highlighting how body movements have changed over time, or how they are different to the larger population.

5.2.5 PHYSICAL SET-UP AND FORM FACTOR

The physical form of a body-tracking system also affects how it can be used. Whether they are mobile or fixed, what size of screens are used, how they are positioned, and the juxtaposition of sensors and displays are all important factors here. For example, gaming systems with large screens, as used in the balance training system described in Chapter 3, encouraged shared viewing and thus social engagement with others in a care home. The down side of large screens is that they tend to be fixed, and this fact combined with the need for interactional space front of the screen means that space needs to be dedicated to these systems, which may be difficult in small homes, for example.

Mobile, more self-contained systems have both benefits and drawbacks. In Assess MS, the design of the physical system needed to be mobile in order that it could be wheeled from one clinical room to another, and so that the system rather than the patient could be moved in order to ensure that the view of the patient was optimal. This added mobility had knock-on consequences in that, without a dedicated space, it was more difficult to understand what the field of view of the camera was in order to avoid obstructing it. In this system, the use of two screens—one for the patient, and another for the clinician was key to solving that problem. In the final version of the system, the clinician's screen was mounted behind the patient's view to encourage the clinician to stand back behind that camera and therefore not get in the way of the field of view. This had knock on consequences for the clinician who could not be entirely sure what the patient was seeing from one moment to the next. While the views shown on both screens were different, the clinician's screen needed to mirror to some extent what the patient was seeing in order that the clinician did not need to constantly check what the patient was viewing.

In the surgical body-tracking systems, some flexibility in the set-up in order to move cameras and screens around depending on the particular procedure to be carried out. What was important here was to fix the screen and camera within easy line of sight of the surgeon so that they could interact without moving from the site of the operation. Also important was to position the camera and display closely together since the surgical team would be oriented toward the images displayed on the screen. The camera had to be well aligned to this orientation to capture gestures as effectively as possible.

5.2.6   SOCIAL SET-UP AND PRACTICES

Finally, it is not enough to design the technology, the movements and the interface. Care and consideration needs to be given to how these systems will be used in an ongoing way and in the wider social context. For example, in lab settings, systems may require considerable effort and expertise when it comes to initiating, calibrating, and maintaining body-tracking systems. For example, markered motion capture systems usually need to be carefully set up and calibrated which can be a time-consuming process requiring considerable expertise.

However, one benefit of lab settings is that equipment can be cordoned off and protected. In real-world settings, body-tracking systems need to be more robust and easy to engage with and maintain if they are to be used by different healthcare professionals and patients. Home settings are perhaps the most challenging in this regard. New systems need to fit within the general routines of the home, be extremely easy to use, and be resilient to the general wear and tear of household life. After all, homes can be shared by many different members of a household (including pets), and so direct "users" are not the only people to consider when designing these systems.

Even so, some systems, such as rehabilitative games, may need to be tailored by a clinician so that they are achievable but also provide some challenge to ensure that ongoing use can then be self-directed. The ability to remotely monitor system use may be helpful here as a way to remotely to check on adherence, but also to signal if there are problems with the system. Another consideration for home use is that they not be obtrusive as many patients are understandably sensitive to the presence of "medicalized" objects in the home.

5.3   THE FUTURE

We are still very much in the early stages of the evolution of body-tracking applications for healthcare, but with the rapidly emerging body of work in this area, we can only assume that progress will be both swift and radically transformative. Body tracking will fundamentally change our ability to assess and diagnose medical conditions as well as support rehabilitation and new kinds of interaction in labs, clinics, surgical suites and at home. This will occur in the form of more accurate and objective ways to measure and monitor the progression of ongoing conditions, which may in time become the new "gold standard" in clinical circles. Body tracking will also offer up entirely new ways to free patients from clinical visits for rehabilitation and diagnosis, enabling them instead to engage in more personal, self-managed healthcare in the course of everyday life. Sensors will also provide new capabilities for remote and unobtrusive monitoring of conditions, and the ability for clinicians to capture more data for assessment in an ongoing way. Body tracking will also enable new possibilities for clinical practice in hospital settings, letting surgeons see into the body in new ways, and radically transform surgical practice. Yet at the same time the future points to benefits not just in terms of the ways in which body movements themselves can be physically assessed, trained, or exploited for interaction, but for all the accompanying emotional and social benefits these will also bring.

Along the way, we have highlighted the many different design challenges to be addressed. But as we build more of these systems, so too will we begin to develop new conventions of use, new guidelines for interface design, a better understanding of the affordances of these new sensors and algorithms, and more evidence of their clinical benefits. Through careful attention to existing work practices, pushing the boundaries of new technology, iterative design, and evaluation, we move the field forward. Through partnerships between the clinical, technical, social science, and design communities, we pool our expertise to get there faster and more effectively.

Bibliography

Alač, M., (2008) Working with Brain Scans: Digital Images and Gestural Interaction in fMRI Laboratory. *Social Studies of Science*, 38(4), pp. 483–508. DOI: 10.1177/0306312708089715. 71

Alankus, G., Lazar, A., May, M., and Kelleher, C. (2010) Toward customizable games for stroke rehabilitation. *Proceedings of CHI '10*, April 10th-15th, Atlanta, GA, pp. 2113–2122. DOI: 10.1145/1753326.1753649. 2, 44, 48, 59, 60

Alexander, N.B. (1994) Postural control in older adults. *Journal of American Geriatric Society*, 42, pp. 93–108. DOI: 10.1111/j.1532-5415.1994.tb06081.x. 58

Allaf, M.E., Jackman, S.V., Schulam, P.G., Cadeddu, J.A., Lee, B.R., Moore, R.G., and Kavoussi, L.R. (1998) Laparoscopic visual field: voice vs. foot pedal interfaces for control of the AESOP robot. *Surgical EndoscopyJournal of American Geriatric Society*, 12(12), pp. 1415–1418, Dec. 1998. DOI: 10.1007/s004649900871. 92

Ashton-James, C.E., Richardson, D.C., de C Williams, A.C., Bianchi-Berthouze, N. and Dekker, P.H. (2014) The impact of pain behaviors on evaluations of warmth and competence. *Pain*, 155(12), pp. 2656–2661. DOI: 10.1016/j.pain.2014.09.031. 57

Aung, M.S.H., Bianchi-Berthouze, N., de C Williams, A.C., and Watson, P. (2014) Automatic Recognition of Fear-Avoidance Behaviour in Chronic Pain Physical Rehabilitation. *Proceedings of PervasiveHealth*, Brussels, Belgium, pp. 151–161. DOI: 10.4108/icst.pervasivehealth.2014.254945. 57

Aung, M.S.H., Romera-Paredes, B., Singh, A., Lim, S., Kanakam, N., de C Williams, A.C., and Bianchi-Berthouze, N. (2013) Getting rid of pain-related behaviour to improve social and self-perception: a technology-based perspective. *Proceedings of WIAMIS'13, 14th International Workshop on Image and Audio Analysis for Multimedia Interactive Service*, July 3rd-5th, Paris, France, pp. 1–4. DOI: 10.1109/wiamis.2013.6616167. 57

Aung, M.S.H., Kaltwang, S., Romera-Paredes, B., Martinez, B., Singh, A., Cella, M., Valstar, M., Meng, H., Kemp, A., Shafizadeh, M., Elkins, A.C., Kanakam, N., de Rothschild, A., Tyler, N., Watson, P.J., de C. Williams, A.C., Pantic, M., and Bianchi-Berthouze, N. (2015) The automatic detection of chronic pain-related expression: Requirements, challenges and the multimodal EmoPain dataset. *IEEE Transactions on Affective Computing*. PP(99). 57

Awad, M., Ferguson, S., and Craig, C. (2014) Designing games for older adults an affordance based approach. *IEEE Serious Games and Applications for Health (SeGAH)*, pp. 1–7. DOI: 10.1109/SeGAH.2014.7067103. 62

Axelrod, L., Fitzpatrick, G., Burridge, J., Mawson, S., Smith, P., Rodden, T., and Ricketts, I. (2009) The reality of homes fit for heroes: design challenges for rehabilitation technology at home. *Journal of Assistive Technologies*, 3(2), pp. 35–43. DOI: 10.1108/17549450200900014. 44, 45

Ayrulu-Erdem, B. and Barshan, B. (2011) Leg motion classification with artificial neural networks using wavelet-based features of gyroscope signals. *Sensors*, 11(2), pp. 1721–1743. DOI: 10.3390/s110201721. 6

Bailenson, J., Patel, K., Nielsen, A., Bajscy, R., Jung, S-H. and Kurillo, G. (2008) The effect of interactivity on learning physical actions in virtual reality. *Media Psychology*, 11(3), pp. 354–376. DOI: 10.1080/15213260802285214. 31

Bainbridge, S., Keeley, B., and Oriel, K. (2011) The effects of the Nintendo Wii Fit on community-dwelling older adults with perceived balance deficits: a pilot study. *Physical & Occupational Therapy in Geriatrics*, 29(2), pp. 126–135. DOI: 10.3109/02703181.2011.569053. 61, 62

Balaam, M., Egglestone, S.R., Fitzpatrick, G., Rodden, T., Hughes, A-M., Wilkinson, A., Nind, T., Axelrod, L., Harris, E., Ricketts, I., Mawson, S., and Burridge, J.(2011) Motivating mobility: Designing for lived motivation in stroke rehabilitation. *Proceedings of CHI '11*, May 7th–12th, Vancouver, BC, Canada, pp. 3073–3082. DOI: 10.1145/1978942.1979397. 44, 45

Barnett, A., Smith, B., Lord, S.R., Williams, M., and Baumand, A. (2003). Community-based group exercise improves balance and reduces falls in at-risk older people: a randomized controlled trial. *Age Ageing*, 32(4), pp. 407–14. DOI: 10.1093/ageing/32.4.407. 59

Barry, G., Galna, B., and Rochester, L. (2014) The role of exergaming in Parkinson's disease rehabilitation: a systematic review of the evidence. *Journal of Neuroengineering and Rehabilitation*, March 7, 11:33. DOI: 10.1186/1743-0003-11-33. 44

Baston, C., Mancini, M., Schoneburg, B., Horak, F., and Rocchi, L. (2014) Postural strategies assessed with inertial sensors in healthy and parkinsonian subjects. *Gait & Posture*, 40(1), pp. 70–75. DOI: 10.1016/j.gaitpost.2014.02.012. 16

Behrens, J., Otte, K., Mansow-Model, S., Brandt, A., and Paul, F. (2014). Kinect-based gait analysis in patients with multiple sclerosis. *Neurology*, 82 (10 Suppl.), P3.135. 16

Bellotti, V., Back, M., Edwards, W. K., Grinter, R. E., Henderson, A., and Lopes, C. (2002) Making sense of sensing systems: five questions for designers and researchers. *Proceedings of CHI '12*, May 5th-10th, Austin, Texas, pp. 415–422. DOI: 10.1145/503376.503450. 97

Bengtsson, S. L., Ullén, F., Ehrsson, H. H., Hashimoto, T., Kito, T., Naito, E., Forssberg, H., and Sadato, N. (2009) Listening to rhythms activates motor and premotor cortices. *Cortex*, 45(1), pp. 62–71. DOI: 10.1016/j.cortex.2008.07.002. 51

Berg, K.O., Wood-Dauphinee, S.L., Williams, J.I., and Maki, B. (1992). Measuring balance in the elderly: validation of an instrument. *Canadian Journal of Public Health*, July-Aug: 83, Suppl 2:S7–11. 62, 65

Betker, A.L., Desai, A., Nett, C., Kapadia, N., and Szturm, T. (2007) Game-based exercises for dynamic short sitting balance rehabilitation of people with chronic spinal cord and traumatic brain injuries. *Physical Therapy*, 87(10), pp. 1389–1398. DOI: 10.2522/ptj.20060229. 61

Bodenheimer, T., Lorig, K., Holman, H., and Grumbach, K. (2002) Patient self-management of chronic disease in primary care. *Journal of American Medical Association*, 288(19), pp. 2469–2475. DOI: 10.1001/jama.288.19.2469. 43

Bonnechère, B., Jansen, B., Salvia, P., Bouzahouene H., Omelina L., Moiseev F., Sholukha V., Cornelis J., Rooze M., and van Sint, J.S. (2014) Validity and reliability of the Kinect within functional assessment activities: Comparison with standard stereophotogrammetry. *Gait & Posture*, 39(1), pp. 593–598. DOI: 10.1016/j.gaitpost.2013.09.018. 5, 15

Bonnechère, B., Sholukha, V., Moiseev, F., Rooze, M., and van Sint, J.S. (2013) From Kinect to anatomically-correct motion modeling: Preliminary results for human application. *Games for Health: Proceedings of the 3rd European Conference on Gaming and Playful Interaction in Health Care*, pp. 15–26. DOI: 10.1007/978-3-658-02897-8_2. 5

Botvinick, M. and Cohen, J. (1998). Rubber hands 'feel' touch that eyes see. *Nature*, 391, pp. 756. DOI: 10.1038/35784. 51

Breivik, H., Collett, B., Ventafridda, V., Cohen, R., and Gallacher, D. (2006) Survey of chronic pain in Europe: prevalence, impact on daily life, and treatment. *European Journal of Pain*, 10(4), pp.287–333. 46

Bushnell, M. C., Čeko, M., and Low, L. A. (2013) Cognitive and emotional control of pain and its disruption in chronic pain. *Nature Neuroscience Review*, 14(1), pp. 502–551. DOI: 10.1038/nrn3516. 46

Cappozzo, A., Catani, F., Della Croce, U., and Leardini, A. (1995). Position and orientation in space of bones during movement: Anatomical frame definition and determination. *Clinical Biomechanics*, 10(4), pp. 171–178. DOI: 10.1016/0268-0033(95)91394-T. 4

Carpintero, E., Perez, C., Morales, R., Garcia, N., Candela, A., and Azorin, J. (2010) Development of a robotic scrub nurse for the operating room. *Proceedings of BioRob 2010, 3rd International Conference on Biomedical Robotics and Biomechatronics*, September 26th-29th, Tokyo, Japan, pp. 504–509. DOI: 10.1109/BIOROB.2010.5626941. 89, 90

CASALA Living Lab. http://www.openlivinglabs.eu/livinglab/casala-living-lab. 42

Centers for Disease Control and Prevention, National Center for Injury Prevention and Control (2013) Web–based injury statistics query and reporting system (WISQARS) [online]. Accessed July, 2014. Centers for Disease Control and Prevention. 58

Cepeda M.S., Carr D.B., and Lau J., Alvarez H. (2006) Music for pain relief. *Clinical Biomechanics Cochrane Database of Systematic Reviews* 2006, Issue 2. DOI: 10.1002/14651858. CD004843.pub2. 51

Chan, J. C., Leung, H., Tang, J. K., and Komura, T. (2011) A virtual reality dance training system using motion capture technology. *IEEE Transactions on Learning Technologies*, 4(2), pp. 187–195. DOI: 10.1109/TLT.2010.27. 31

Charbonneau, E., Miller, A., Wingrave, C., and LaViola, J. (2009) Understanding visual interfaces for next generation of dance-based rhythm video games. *Proceedings of Sandbox '09 ACM SIGGRAPH Symposium on Video Games*, August 3rd-7th, New Orleans, Louisiana, pp. 119–126. DOI: 10.1145/1581073.1581092. 31

Chez, C., Rikakis, T., Dubois, R. L., and Cook, R. (2000) An auditory display system for aiding interjoint coordination. *Proceedings International Conference on Auditory Display*, April 2nd-5th, Atlanta, GA. 51

Chua, P.T., Crivella, R., Daly, B., Hu,, N., Schaaf, R., Ventura, D., Camill, T., Hodgins, J., and Pausch. P. (2003) Training for physical tasks in virtual environments: Tai Chi. *Proceedings of the IEEE Virtual Reality 2003 (VR '03)*, Washington, DC, pp. 87–94. DOI: 10.1109/VR.2003.1191125. 31

Clark R.A., Bryant A.L., Pua Y., McCrory P., Bennell K., and Hunt M. (2010) Validity and reliability of the Nintendo Wii Balance Board for assessment of standing balance. *Gait & Posture*, 31(3), pp. 307–310. DOI: 10.1016/j.gaitpost.2009.11.012. 62

Clark, R. A., Pua, Y.-H., Fortin, K., Ritchie, C., Webster, K.E., Denehy, L., and Bryant, A.L. (2012) Validity of the Microsoft Kinect for assessment of postural control. *Gait & Posture* 36(3), pp. 372–377. DOI: 10.1016/j.gaitpost.2012.03.033. 15

Cohen, J. A., Reingold, S. C., Polman, C. H., and Wolinsky, J. S. (2012) Disability outcome measures in multiple sclerosis clinical trials: current status and future prospects. *Lancet Neurology*,11(5), pp. 467–76. DOI: 10.1016/S1474-4422(12)70059-5. 15

Crea, S., Donati, M., de Rossi, M.S., Oddo, C. M., and Vitiello, N. (2014) A wireless flexible sensorized insole for gait analysis. *Sensors*, 14(1), pp. 1073–1093. DOI: 10.3390/s140101073. 6

Crombez, G., Eccleston, C., Van Damme, S., Vlaeyen, J.W.S., and Karoly, P. (2012) Fear-avoidance model of chronic pain: the next generation. *Clinical Journal of Pain*, 28(6), pp. 475–483. DOI: 10.1097/AJP.0b013e3182385392. 47

de Rossi, D. Lorussi, F., Mazzoldi, A., Scilingo, E.P. ,and Orsini, P. (2001) Active dressware: Wearable proprioceptive systems based on electroactive polymers. *Proceedings of ISWC 2001, 5th International Symposium on Wearable Computers*, 8th-9th October, Zurich, Switzerland, p. 161. DOI: 10.1109/iswc.2001.962123. 6

de Vignemont, F., Ehrsson, H. H., and Haggard, P. (2005). Bodily illusions modulate tactile perception. *Current Biology*, 15(14), pp. 1286–1290. DOI: 10.1016/j.cub.2005.06.067. 51

Dobson, F., Morris, M.E., Baker, R., and Graham, H.K. (2007) Gait classification in children with cerebral palsy: A systematic review. *Gait & Posture*, 25(1), pp. 140–152. DOI: 10.1016/j.gaitpost.2006.01.003. 1

Douglas, M. (1966) *Purity and Danger*. London: Routledge. DOI: 10.4324/9780203361832.

Du, S., Yuan, C., Xiao, X., Chu, J., Qiu, Y., and Qian, H. (2011) Self-management programs for chronic musculoskeletal pain conditions: A systematic review and meta-analysis. *Patient Education and Counseling*, 85(3), e299–e310. DOI: 10.1016/j.pec.2011.02.021. 43

Dubus, G. (2012) Evaluation of four models for the sonification of elite rowing. *Journal of Multimodal User Interfaces*, 5(3-4), pp. 143–156. DOI: 10.1007/s12193-011-0085-1. 51

Ebert, L., Hatch, G., Ampanozi, G., Thali, M., and Ross, S. (2011) You can't touch this: Touch-free navigation through radiological images. *Surgical Innovation*, 19(3), pp. 301–307. DOI: 10.1177/1553350611425508. 77

Ebert, L., Hatch, G., Ampanozi, G., Thali, M., and Ross, S. (2012) Virtopsy - Kinect Gesture Control Plugin for OsiriX. http://www.youtube.com/watch?v=wyeQi-0yC_w. 77

Effenberg, A.O., Fehse, U., and Weber, A. (2011) Movement sonification: Audiovisual benefits on motor learning. *BIO Web of Conferences*, 1, 00022. DOI: 10.1051/bioconf/20110100022. 51

Effenberg, A.O., Weber, A., Mattes, K., Fehse, U., and Mechling, H. (2007) Motor learning and auditory information: Is movement sonification efficient? *Journal of Sport & Exercise Psychology*, 29, p.S69. 51

Effenberg, A.O. (2005) Movement sonification: Effects on perception and action. *IEEE Multimedia*, 12(2), pp. 53–59. DOI: 10.1109/MMUL.2005.31. 51

Ekman, P. and Friesen, W. (1978) *Facial Action Coding System: A Technique for the Measurement of Facial Movement*, Palo Alto, CA: Consulting Psychologists Press. 56

El-Gohary, M., Pearson, S., McNames, J., Mancini, M., Horak, F., Mellone, S., and Chiari, L. (2014) Continuous monitoring of turning in patients with movement disability. *Sensors*, 14(1), pp. 356–369. DOI: 10.3390/s140100356. 16

Fahn, S., Elton, R.L., and UPDRS committee (1987) Unified Parkinson's disease rating scale. In Fahn, S., Marsden, C. D., Goldstein, M. and Calne, D.B. (Eds) *Recent developments in Parkinson's Disease*. Florham Park, NJ: McMillan Health Care, pp. 153–163. 14

Faivre, A., Dahan, M., Parratte, B., and Monnier, G. (2004) Instrumented shoes for pathological gait assessment. *Mechanics Research Communications*, 31(5), pp. 627–632. DOI: 10.1016/j. mechrescom.2003.10.008. 7

Ferrarin, M., Rabuffetti, M., Bacchini, M, Casiraghi, A., Castagna, A., Pizzi, A., and Montesano, A. (2014) Does gait analysis change clinical decision-making in post-stroke patients? Results from a pragmatic prospective observational study. *European Journal of Physical and Rehabilitation Medicine*, 51(2), pp. 171–84. 2

Finlay, P.A. (2001) A robotic camera holder for laparoscopy. *Proceedings of ICAR '01, 10th International Conference on Advanced Robotics, Workshop 2 on Medical Robotics*, August Budapest, Hungary, pp. 129–132. 92

Finlay, P.A. and Ornstein, M.H. (1995) Controlling the movement of a surgical laparoscope. *IEEE Engineering in Medicine and Biology Magazine*, 14(3), pp. 289–291. DOI: 10.1109/51.391775. 92

Fitzpatrick, G., Balaam, M., Axelrod, L., Harris, E., McAllister, G., Hughes, A-M., Burridge, J., Nind, T., Ricketts, I., Wilkinson, A., Mawson, S., Rennick-Egglestone, S., Rodden, T., Probert Smith, P., Shublaq, N., and Robertson, Z. (2010) Designing for rehabilitation at home. *CHI Workshop on Interactive Systems in Healthcare*, 10th April, Atlanta, GA, pp. 49–52. 45

Frees, S., Kessler, G.D., and Kay, E. (2007) PRISM Interaction for Enhancing Control in Immersive Virtual Environments. *Transactions on Computer Human Interaction (ToCHI)*, 14(1), Article 2, May, 2007. DOI: 10.1145/1229855.1229857. 87

Gabel, M., Renshaw, E. Schuster, A., and Gilad-Bachrach, R. (2012) Full body gait analysis with kinect. *Proceedings of EMBC '12, International Conference of the Engineering in Medicine and Biology Society*, August 28th – September 1st, San Diego, CA, pp. 1964–1967. DOI: 10.1109/embc.2012.6346340. 5, 16

Gallo, L., Placitelli, A.P., and Ciampi, M. (2011) Controller-free exploration of medical image data: experiencing the Kinect. *Proceedings of CBMS 2011, International Symposium on Computer-Based Medical Systems*, June 27th-30th, Bristol, UK, pp. 1–6. DOI: 10.1109/CBMS.2011.5999138. 77, 88

Gatchel, R.J., Peng, Y.B., Peters, M.L., Fuchs, P.N., and Turk, D.C. (2007) The biopsychosocial approach to chronic pain: scientific advances and future directions. *Psychological Bulletin* 133(4), pp. 581–624. DOI: 10.1037/0033-2909.133.4.581. 46

Geisser, M.E., Haig, A.J., Wallbom, A.S., and Wiggert, E.A. (2004). Pain-related fear, lumbar flexion, and dynamic EMG among persons with chronic musculoskeletal low back pain. *Clinical Journal of Pain*, 20(2), pp. 61–69. DOI: 10.1097/00002508-200403000-00001. 47

Gerling, K. and Masuch, M. (2011) When gaming is not suitable for everyone: Playtesting Wii games with frail elderly. *Workshop on Game Accessibility: Xtreme Interaction Design (GAXID'11)*, 28th June, Bordeaux, France. 62

Gerling, K., Mandryk, R., and Linehan, C. (2015) Long-term use of motion-based video games in care home settings. *Proceedings of CHI '15*, April 18th-23rd, Seoul, Korea, pp. 1573–1582. DOI: 10.1145/2702123.2702125. 66, 67

Gerling, K.M., Livingston, I.J., Nacke, L.E., and Mandryk, R.L. (2012) Full-body motion-based game interaction for older adults. *Proceedings of CHI '12*, May 5th-10th, Austin, Texas, pp. 1873–1882. DOI: 10.1145/2207676.2208324. 44

Geurts, L., Abeele, V.V., Husson, J., Windey, F. van Overveldt, M., Annema, J.H., and Desmet, S. (2011) Digital games for physical therapy: fulfilling the need for calibration and adaptation. *Proceedings of TEI '11, 5th International Conference on Tangible, Embedded, and Embodied Interaction*, January 23rd-26th, Funchal, Madeira, Portugal, pp. 117–124. DOI: 10.1145/1935701.1935725. 2, 48, 59

Gladstone, D. J., Danells, C. J., and Black, S. E. (2002) The Fugl-Meyer assessment of motor recovery after stroke: A critical review of its measurement properties. *Neurorehabilitation and Neural Repair*, 16(3), pp. 232–240. DOI: 10.1177/154596802401105171. 15

Gomez, J. B., Ceballos, A., Prieto, F., and Redarce, T. (2009) Mouth gesture and voice command based robot command interface. *Proceedings of ICRA '09, International Conference on Robotics and Automation*, May 12th-17th, Kobe, Japan, pp. 333–338. DOI: 10.1109/robot.2009.5152858. 92

Goodwin, C. (1994) Professional vision. *American Anthropologist*, 96(3), pp. 606–633. DOI: 10.1525/aa.1994.96.3.02a00100. 8, 71

Goodwin, C. (2000) Practices of seeing: Visual analysis: An ethnomethodological approach. In van Leeuwen, T. and Carey Jewit, C. (Eds) *Handbook of Visual Analysis*. London: Sage Publications. 8, 71, 93

Graetzel, C., Fong, T., Grange, S., and Baur, C. (2004) A non-contact mouse for surgeon-computer interaction. *Technology and Health Care*, 12(3), pp. 245–57. 73, 75

Graham, D.L., Ferreira, H.A., and Freitas, P.P. (2004) Magnetoresistive-based biosensors and bio-chips. *Trends in Biotechnology*, 22(9), pp. 455–462. DOI: 10.1016/j.tibtech.2004.06.006. 6

Greatbatch, D., Luff, P., Heath, C., and Campion, P. (1993) Interpersonal communication and human-computer interaction: an examination of the use of computers in medical consultations. *Interacting with Computers*, 5(2), pp. 193–216. DOI: 10.1016/0953-5438(93)90018-O. 23, 24

Gromala, D., Song, M., Yim, J.D., Fox, T., Barnes, S.J., Nazemi, M., Shaw, C., and Squire, P. (2011) Immersive VR: a non-pharmacological analgesic for chronic pain? *Extended Abstracts of CHI '11*, May 7th-12th, Vancouver, Canada, pp. 1171–1176. DOI: 10.1145/1979742.1979704. 49

Grönvall, E. and Lundberg, S. (2014) On challenges designing the home as a place for care. In Holzinger, A., Ziefle, M. and Röcker, C. (eds.) *Pervasive Health*. London: Springer-Verlag, pp. 19–45. DOI: 10.1007/978-1-4471-6413-5_2. 44, 45

Hamaoui, A., Do, M.C., Poupard, L., and Bouisset, S. (2002) Does respiration perturb body balance more in chronic low back pain subjects than in healthy subjects? *Clinical Biomechanics*, 17(7), pp. 548–550. DOI: 10.1016/S0268-0033(02)00042-6. 47

Hamill, J. and Knutzen, K. (2003) *Biomechanical Basis of Human Movement*. Baltimore: Lippincott, Williams & Wilkins. 18

Hammal, Z. and Cohn, J.F. (2012). Automatic detection of pain intensity. *Proceedings of ICMI '12, International Conference on Multimodal Interaction*, October 22nd–26th, Santa Monica, CA, pp. 47–52. DOI: 10.1145/2388676.2388688. 56

Harley, D., Fitzpatrick, G., Axelrod, L., White, G., and McAllister, G. (2010) Making the Wii at home: game play by older people in sheltered housing. *Proceedings of USAB'10, International Conference on HCI in Work and Learning, Life and Leisure*, November 4th-5th, Klagenfurt, Austria, pp. 156–176. DOI: 10.1007/978-3-642-16607-5_10. 66

Hartswood, M., Procter, R., Rouncefield, M., Slack, R. Soutter, J., and Voss, A. (2003) Repairing the machine: A case study of the evaluation of computer-aided detection tools in breast screening. *Proceedings of ECSCW '03, 8th European Conference on Computer-Supported Cooperative Work,* September 14th -18th, Helsinki, Finland, pp. 375–394. DOI: 10.1007/978-94-010-0068-0_20. 8, 14, 20

Hass, C., Waddell, D., Fleming, R.P., Juncos, J.L., and Gregor, R.J. (2005) Gait initiation and dynamic balance control in Parkinson's disease. *Archives of Physical Medicine and Rehabilitation*, 86(11), pp. 2172–2176. DOI: 10.1016/j.apmr.2005.05.013. 1

Heath, C. (2003) *Body Movement and Speech in Medical Interaction.* Cambridge: Cambridge University Press. 23, 24

Herlocker, J.L., Konstan, J., and Riedl, J. (2000) Explaining collaborative filtering recommendations. *Proceedings of CSCW 2000*, 2nd-6th December, Philadelphoa, PA, pp. 241–250. DOI: 10.1145/358916.358995. 37

Hirsche, R., Williams, B., Jones, A., and Manns, P. (2011) Chronic disease self-management for individuals with stroke, multiple sclerosis and spinal cord injury. *Disability and Rehabilitation*, 33(13-14), pp. 1136–1146. DOI: 10.3109/09638288.2010.523103. 43

Hobart, J.C., Cano, S.J., Zajicek, J.P., and Thompson, A.J. (2007) Rating scales as outcome measures for clinical trials in neurology: problems, solutions, and recommendations. *Lancet Neurology* 6(12), pp. 1094–105. DOI: 10.1016/S1474-4422(07)70290-9. 14, 15, 18

Hoff, J.I., van den Plas, A.A., Wagemans, E.A., and van Hilten, B.J. (2001a) Accelerometric assessment of levodopa-induced dyskinesias in Parkinson's disease. *Movement Disorders*, 16(1), pp. 58–61. DOI: 10.1002/1531-8257(200101)16:1<58::AID-MDS1018>3.0.CO;2-9. 16

Hoff, J.I., Wagemans, E.A., and van Hilten, B.J. (2001b) Ambulatory objective assessment of tremor in Parkinson's disease. *Clinical Neuropharmacology* 24(5), pp. 280–3. DOI: 10.1097/00002826-200109000-00004. 16

Horak, F.B. and Mancini, M. (2013) Objective biomarkers of balance and gait for Parkinson's disease using body worn sensors. *Movement Disorders*, 28(11), pp 1544–1551. DOI: 10.1002/mds.25684. 16

Jacob, M., Li, Y., Akingba, G., and Wachs, J.P. (2012) Gestonurse: a robotic surgical nurse for handling surgical instruments in the operating room. *Journal of Robotic Surgery*, 6(1), pp. 53–63. DOI: 10.1007/s11701-011-0325-0. 89, 90

Jacob, M.G., Li, Y.T., and Wachs, J.P. (2011) A gesture driven robotic scrub nurse. *Proceedings of IEEE International Conference on Systems, Man, and Cybernetics,* October 9th-12th, Anchorage, AK, pp. 2039–2044. DOI: 10.1109/icsmc.2011.6083972. 89, 90

Jansen-Kosterink, S.M., Huis In 't Veld, R.M., Schönauer, C., Kaufmann, H., Hermens, H.J., and Vollenbroek-Hutten, M.M.R. (2013) A Serious exergame for patients suffering from chronic musculoskeletal back and neck pain: A pilot study. *Games for Health Journal,* 2(5), pp. 299–307. DOI: 10.1089/g4h.2013.0043. 48

Janssen, W.G., Bussmann, H.B., and Stam, H.J. (2002) Determinants of the sit-to-stand movement: A review. *Physical Therapy,* 82(9), pp. 866–879.

Jaume-i-Capó, A., Moyà-Alcover, B. and Varona, J. (2014) Design issues for vision-based motor-rehabilitation serious games. In Brooks, A.L., Brahnam, S. and Jain, L.C. (Eds.) *Technologies of Inclusive Well-Being.* Berlin: Springer-Verlag, pp. 13–24. DOI: 10.1007/978-3-642-45432-5_2. 2

Johnson, R., O'Hara, K., Sellen, A., Cousins, C., and Criminisi, A. (2011) Exploring the potential for touchless interaction in image-guided interventional radiology. *Proceedings of CHI '11,* 7th-12th May, Vancouver, Canada, pp. 3323–3332. DOI: 10.1145/1978942.1979436. 72, 73, 74, 78, 79, 87

Jung, Y., Li, J.K., Janissa, N.S., Gladys, W.L.C., and Lee, K.M. (2009) Games for a better life: effects of playing Wii games on the well-being of seniors in a long-term care facility. *Proceedings of IE '09,* ACM Press, article 5. DOI: 10.1145/1746050.1746055. 66

Kaltwang, S., Rudovic, O., and Pantic, M. (2012) Continuous pain intensity estimation from facial expressions. *Advances in Visual Computing,* LCNS, vol. 7432, pp. 368–377. DOI: 10.1007/978-3-642-33191-6_36. 56

Kamm, C.P., Uitdehaag, B. M., and Polman, C.H. (2014) Multiple sclerosis: current knowledge and future outlook. *European Neurology,* 72(3-4), pp. 132–141. DOI: 10.1159/000360528. 18

Katz, P. (1981) Ritual in the operating room. *Ethnology,* 20(4), pp. 335–50. DOI: 10.2307/3773355. 72

Keefe, F.J. and Block, A.R. (1982). Development of an observation method for assessing pain behaviour in chronic low back pain patients. *Behaviour Therapy,* 13(4), pp. 363–375. DOI: 10.1016/S0005-7894(82)80001-4. 49

Kenyon, G.P. and Thaut, M.H. (2005) Rhythmic-drive optimization of motor control. In M. H. Thaut (Ed) *Rhythm, Music and the Brain: Scientific Foundations and Clinical Applications* (pp. 85-112). New York: Routledge Chapman & Hall. 51

Kerns, R.D., Sellinger, J., and Goodin, B.J. (2011). Psychological treatment of chronic pain. *Annual Review of Clinical Psychology*, 7, pp. 411–434. DOI: 10.1146/annurev-clin-psy-090310-120430. 46, 69

Kiani, K., Snijders, C., and Gelsema, E. (1997) Computerized analysis of daily life motor activity for ambulatory monitoring. *Technology and Health Care*, 5(4), pp. 307–18. 16

Kim, J.W., Lee, J.H., Kwon, Y., Kim, C.S., Eom, G.M., Koh, S.B., Kwon, D.Y., and Park, K.W. (2011) Quantification of bradykinesia during clinical finger taps using a gyrosensor in patients with Parkinson's disease. *Medical and Biological Engineering and Computing*, 49(3), pp. 365–371. DOI: 10.1007/s11517-010-0697-8. 16

Kipshagen, T., Graw, M., Tronnier, V., Bonsanto, M., and Hofmann, U. (2009) Touch- and mark-er-free interaction with medical software. *Proceedings of the World Congress on Medical Physics and Biomedical Engineering*, September 7th-12th, Munich, Germany, pp. 75–78. DOI: 10.1007/978-3-642-03906-5_21. 76

Kleiman-Weiner, M. and Berger J. (2006). The sound of one arm swinging: a model for multi-dimensional auditory display of physical motion. *Proceedings of the 12th International Conference on Auditory Display*, May 18th-22nd, Copenhagen, Denmark, pp. 278–280. 51

Kochan, A. (2005) Scalpel please, robot: Penelope's debut in the operating room. *Industrial Robot*, 32(6), pp. 449–451. DOI: 10.1108/01439910510629136. 89, 90

König, W.A., Gerken, J., Dierdorf, S., and Reiterer, H. (2009) Adaptive pointing–design and eval-uation of a precision enhancing technique for absolute pointing devices. *Proceedings of INTERACT '09, 12th IFIP TC 13 International Conference*, August 24th-28th, Uppsala, Sweden, pp. 658–671. DOI: 10.1007/978-3-642-03655-2_73. 88

Kopper, R., Bowman, D., Silva, M.G., and McMahan, R.P. (2010) A human motor behavior model for distal pointing tasks. *International Journal of Human-Computer Studies*, 68(10), pp. 603–615. DOI: 10.1016/j.ijhcs.2010.05.001. 87

Kuan, T.S., Tsou, J.Y., and Su, F.C. (1999) Hemiplegic gait of stroke patients: the effect of using a cane. *Archives of Physical Medicine and Rehabilitation*, 80(7), pp. 777–84. DOI: 10.1016/S0003-9993(99)90227-7. 2

Kulesza, T., Stumpf, S., Burnett, M., and Kwan, I. (2012) Tell me more?: the effects of mental model soundness on personalizing an intelligent agent. *Proceedings of CHI '12*, Austin, Texas, pp. 1–10. DOI: 10.1145/2207676.2207678. 37

Kulesza, T., Stumpf, S., Wong, W-K., Burnett, M., Perona, S., Ko, A., and Oerst, I. (2011) Why-ori-ented end-user debugging of naive Bayes text classification. *ACM Transactions on Interac-tive Intelligent Systems*, 1(1), pp. 1–31. DOI: 10.1145/2030365.2030367. 37

Kurtzke, J.F. (1983) Rating neurologic impairment in multiple sclerosis: An expanded disability status scale (EDSS). *Neurology*, 33(11), pp. 1444–1452. DOI: 10.1212/WNL.33.11.1444. 14, 18

Laughton, C.A., Slavin, M., Katdare, K., Nolan, L., Bean, J.F., Kerrigan, D.C., Phillips, E., Lipsitz, L. A., and Collins, J.J. (2003). Aging, muscle activity, and balance control: Physiological changes associated with balance impairment. *Gait & Posture*, 18(2), pp101–108. DOI: 10.1016/S0966-6362(02)00200-X. 58

Laver, K., Ratcliffe, J., George, S., Burgess, L., and Crotty, M. (2011) Is the Nintendo Wii Fit really acceptable to older people? A discrete choice experiment. *BMC Geriatrics*, 2011, 11:64. DOI: 10.1186/1471-2318-11-64. 62

Leddy, A.L., Crowner, B.E., and Earhart, G.M. (2011) Utility of the Mini-BESTest, BESTest, and BESTest sections for balance assessments in individuals with Parkinson disease. *Journal of Neurologic Physical Therapy*, 35(2), pp. 90–97. DOI: 10.1097/NPT.0b013e31821a620c. 15

Lee, A.S., Cholewicki, J., Reeves, N.P. ,and Zazulak, B.T. (2010) Comparison of trunk proprioception between patients with low back pain and healthy controls. *Archives of Physical Medicine and Rehabilitation*, 91(9), pp. 1327–1331. DOI: 10.1016/j.apmr.2010.06.004. 47, 51

Legrain, V., Damme, S.V., Eccleston, C., Davis, K.D., Seminowicz, D.A., and Crombez, G. (2009) A neurocognitive model of attention to pain: behavioral and neuroimaging evidence. *Pain*, 144(3), pp. 230–232. DOI: 10.1016/j.pain.2009.03.020. 46

Levac, D., Miller, P., and Missiuna, C. (2012) Usual and virtual reality video game-based physiotherapy for children and youth with acquired brain injuries. *Physical & Occupational Therapy in Pediatrics*, 32(2), pp. 180–195. DOI: 10.3109/01942638.2011.616266. 59

Levin, M.F., Sveistrup, H., and Subramanian, S.K. (2010) Feedback and virtual environments for motor learning and rehabilitation. *Schedae*, 1, pp. 19–36. 68

Lopez-Meyer, P., Fulk, G.D., and Sazonov, E.S. (2011) Automatic detection of temporal gait parameters in post stroke individuals. *IEEE Transactions on Information Technology in Biomedicine*, 15(4), pp. 594–600. DOI: 10.1109/TITB.2011.2112773. 2

Lord, S.R., Ward, J.A., Williams, P., and Anstey, K. (1994) Physiological factors associated with falls in older community-dwelling women. *Journal of the American Geriatrics Society*, 42(10), pp. 1110–1117. DOI: 10.1111/j.1532-5415.1994.tb06218.x. 58

Lowes, L.P., Alfano, L.N., Yetter, B.A., Worthen-Chaudhari, L., Hinchman, W., Savage, J., Samona, P., Flanian, K.M., and Mendell, J.R. (2013) Proof of concept of the ability of the kinect to quantify upper extremity function in dystrophinopathy. *Public Library of Science Currents*, 5, pii. DOI: 10.1371/currents.md.9ab5d872bbb944c6035c9f9bfd314ee2. 16

Lucy, P., Cohn, J.F., Prkachin, K.M., Solomon, P.E., and Matthews, I. (2011) Painful Data: The UNBC-McMaster shoulder pain expression archive database. *Proceedings of FG 2011, International Conference on Face and Gesture Recognition*, March 21st-25th, Santa Barbara, CA, , pp. 57–64. DOI: 10.1109/fg.2011.5771462. 56

Lynch, M. (1985) Discipline and the material form of images: An analysis of scientific visibility. *Social Studies of Science*, 15(1), pp. 37–66. DOI: 10.1177/030631285015001002. 8, 9

Lynch, M. (1990a) The externalized retina: Selection and mathematization in the visual documentation of objects in the life sciences. In Lynch, M. and Woolgar, S. (Eds) *Representation in Scientific Practice*, pp. 153–186. Cambridge MA: MIT Press. 8, 71

Lynch, M. (1990b) Drawing things together. In Lynch, M. and Woolgar, S. (Eds.) *Representation in a Scientific Practice*, pp. 19-68. Cambridge, MA: MIT Press. 8, 71

Maki, B.E., Holliday, P.J., and Topper, A.K. (1991) Fear of falling and postural performance in the elderly. *Journal of Gerontology*, 46(4), pp. 123–131. DOI: 10.1093/geronj/46.4.M123. 58

Maki, B.E., Holliday, P.J., and Topper, A. K. (1994) A prospective study of postural balance and risk of falling in an ambulatory and independent elderly population. *Journal of Gerontology*, 49(2), pp. M72–84. DOI: 10.1093/geronj/49.2.M72. 58

Marey, E-J. (1890) *Physiologie du Movement: Le Vol des Oiseaux*. Paris: G. Masson. 1

Marston, H.R., Greenlay, S., and van Hoof, J. (2013) Understanding the Nintendo Wii and Microsoft Kinect consoles in long-term care facilities. *Technology and Disability*, 25(2), pp. 77–85. DOI: 10.3233/TAD-130369. 66, 67

Martel, M.O., Wideman, T.H., and Sullivan, M.J.L. (2012) Patients who display protective pain behaviors are viewed as less likable, less dependable, and less likely to return to work. *Pain*, 153(4), pp. 843–849. DOI: 10.1016/j.pain.2012.01.007. 57

Mathiowetz, V., Weber, K., Kashman, N., and Volland, G. (1985) Adult norms for the nine hole peg test of finger dexterity. *Occupational Therapy Journal of Research: Occupation, Participation and Health*, 5(1), pp. 24–38. DOI: 10.1177/153944928500500102. 18, 24

Matsubara, M., Kadone, H., Iguchi, M., Terasawa, H., and Suzuki, K. (2013). The effectiveness of auditory biofeedback on a tracking task for ankle joint movements in rehabilitation. *Proceedings of Interactive Sonification (ISON) Workshop*, December 10th, Erlangen, Germany. 51

Mazilu, S., Blanke, U., Hardegger, M., Tröster, G., Gazit, E., and Hausdorff, J.M. (2014) GaitAssist: a daily-life support and training system for parkinson's disease patients with freezing of gait. *Proceedings of CHI '14*, Toronto, Canada, pp. 2531–2540. DOI: 10.1145/2556288.2557278. 16

Mazzoldi, A., de Rossi, D. Lorussi, F., Scilingo, E.P., and Paradiso, R. (2002) Smart textiles for wearable motion capture systems. *AUTEX Research Journal*, 2, pp. 199–204. 6

Mentiplay, B.F., Clark, R.A., Mullins, A., Bryant, A.L., Bartold, S., and Paterson, K. (2013) Reliability and validity of the Microsoft Kinect for evaluating static foot posture. *Journal of Foot and Ankle Research*, 6: 14. DOI: 10.1186/1757-1146-6-14. 16

Mentis, H., O'Hara, K., Sellen, A., and Trivedi, R. (2012) Interaction proxemics and image use in neurosurgery. *Proceedings of CHI '12*, 5th-10th May, Austin, Texas, pp. 927–936. DOI: 10.1145/2207676.2208536. 72, 73, 78

Mentis, H.M. and Taylor, A. (2013) Imaging the body: Embodied vision in minimally invasive surgery. *Proceedings of CHI '13, Conference Human Factors in Computing Systems*, 27th April- 2nd May, Paris, France, pp. 1479–1488. DOI: 10.1145/2470654.2466197. 8, 9

Mentis, H.M., Shewbridge, R., Powell, S., Armstrong, M., Fishman, P., and Shulman, L. (2016) Co-interpreting movement with sensors: Assessing Parkinson's patients' deep brain stimulation programming. In *Human-Computer Interaction: Special Issue on Body Tracking and Healthcare*. 23

Mentis, H.M., Shewbridge, R., Powell, S., Fishman, P., and Shulman, L. (2015) Being seen: Co-interpreting Parkinson's patient's movement ability in deep brain stimulation programming. *Proceedings of CHI '15 Conference on Human Factors in Computing Systems*, April 18th-23rd, Seoul, Korea, pp. 511–520. DOI: 10.1145/2702123.2702342. 23

Mithun, J., Cange, C., Packer, R., and Wachs, J.P. (2012) Intention, context and gesture recognition for sterile MRI navigation in the operating room. *Lecture Notes in Computer Science*, 7441, pp. 220–227. 84

Miyawaki, F., Masamune, K., Suzuki, S., Yoshimitsu, K., and Vain, J. (2005) Scrub nurse robot system - intraoperative motion analysis of a scrub nurse and timed-automata-based model for surgery. *IEEE Transactions on Industrial Electronics*, 52(5), pp. 1227–1235. DOI: 10.1109/TIE.2005.855692. 89

Morris, J.R.W. (1973) Accelerometry—A technique for the measurement of human body movements. *Journal of Biomechanics*, 6(6), pp. 729–736. DOI: 10.1016/0021-9290(73)90029-8. 6

Morris, M.E., Matyas, T.A., Iansek, R., and Summers, J.J. (1996) Temporal stability of gait in Parkinson's disease. *Physical Therapy*, 76(7), pp. 763–777. 2

Morrison, C., Huckvale, K., Corish, R., Dorn, J., Kontschieder, P., O'Hara, K., Assess MS Team, Criminisi, A., and Sellen, A. (2016) Assessing multiple sclerosis with Kinect: Designing a machine-learning-based system for use in the real world. *Human-Computer Interaction:*

Special Issue on Body Tracking and Healthcare. DOI: 10.1080/07370024.2015.1093421. 16, 17, 20, 25

Mouawad, M., Doust, C., Max, M., and McNulty, P., (2011) Wii-based movement therapy to promote improved upper extremity function post-stroke: A pilot study. *Journal of Rehabilitation Medicine*, 43(6), pp. 527–533. DOI: 10.2340/16501977-0816. 44

Muybridge, E. (1893) *Descriptive Zoopraxography: The Science of Animal Location Made Popular.* Chicago: Lakeside Press. 1

Mynatt, E.D., Abowd, G.D., Mamykina, L. ,and Kientz, J.A. (2010) Understanding the potential of ubiquitous computing for chronic disease management. In Hayes, B.M. and Aspray, W. (Eds.) *Health Informatics: A Patient-Centered Approach to Diabetes*, Cambridge: MIT Press. DOI: 10.7551/mitpress/9780262014328.003.0003. 46

Nishikawa, A., Hosoi, T., Koara, K., Negoro, D., Hikita, A., Asano, S., Kakutani, H., Miyazaki, F., Sekimoto, M., Yasui, M., Miyake, Y., Takiguchi, S., and Monden, M. (2003) FAce MOUSe: A novel human-machine interface for controlling the position of a laparoscope. *IEEE Transactions on Robotics and Automation*, 19(5), pp. 825–841. DOI: 10.1109/TRA.2003.817093. 83, 92

Nomm, S., Petlenkov, E., Vain, J., Belikov, J. Miyawaki, F., and Yoshimitsu, K. (2008) Recognition of the surgeon's motions during endoscopic operation by statistics based algorithm and neural networks based ANARX models. *Proceedings of International Federation of Automatic Control World Congress*, COEX South Korea, pp. 14773–14778. 89

Nomm, S., Petlenkov, E., Vain, J., Yoshimitsu, K., Ohnuma, K., and Sadahiro, T. (2007) Nn-based anarx model of the surgeon's hand for the motion recognition. *Proceedings of 4th COE Workshop on Human Adaptive Mechatronics (HAM)*, pp. 19–24. 89

Nyman, S.R. and Victor, C.R. (2011) Older people's participation in and engagement with falls prevention interventions in community settings: An augment to the Cochrane systematic review. *Age and Ageing*, 41(1), pp. 16–23. DOI: 10.1093/ageing/afr103. 59

O'Hara, K., Gonzalez, G., Penney, G., Sellen, A. Corish, R., Mentis, H., Varnavas, A., Criminisi, A., Rouncefield, M., Dastur, N. ,and Carrell, T. (2014a) Interactional order and constructed ways of seeing with touchless imaging systems in surgery. *Journal of Computer Supported Cooperative Work*, 23(3), pp. 299–337. DOI: 10.1007/s10606-014-9203-4. 8, 17, 78, 87

O'Hara, K., Gonzalez, G., Sellen, A., Penney, G., Varnavas, A., Mentis, H. Criminisi, A., Corish, R., Rouncefield, M., Dastur, N., and Carrell, T. (2014b) Touchless interaction in surgery. *Communications of the ACM*, 57(1), pp. 70–77. DOI: 10.1145/2541883.2541899. 17, 78

O'Hara, K., Harper, R., Mentis, H. Sellen, A., and Taylor, A. (2013) On the naturalness of touchless: putting the "Interaction" back into NUI. *Transactions on Computer-Human Interaction*, 20(1), Article 5. DOI: 10.1145/2442106.2442111. 72, 73, 74, 78, 79, 93

Ohnuma, K., Masamune, K., Yoshimitsu, K., Vain, J., Fukui, Y., and Miyawaki, F., (2006) Analysis and recognition of a surgeon's motions in laparoscopic cholecystectomy giving a scrub nurse robot suitable timings for instrument exchange. *Proceedings of the 3rd COE Workshop on Human Adaptive Mechatronics (HAM)*, pp. 219–223. 89

Omelina, L., Jansen, B., Bonnechère, B., Van Sint Jan, S., and Cornelis, J. (2012) Serious games for physical rehabilitation: designing highly configurable and adaptable games. *Proceedings of ICDVRAT'12, 9th International Conference on Disability, Virtual Reality & Associated Technologies*, Laval, France, pp. 195–201. 48

Palisano, R. Rosenbloom, P. , Bartlett, D., and Livingstone, M.H. (2008) Content validity of the expanded and revised gross motor function classification system. *Developmental Medicine and Child Neurology*, 50(10), pp. 744–750. DOI: 10.1111/j.1469-8749.2008.03089.x. 14

Palisano, R. Rosenbloom, P., Walter, S., Russell, D., Wood, E., and Galuppi, B. (1997) Development and reliability of a system to classify gross motor function in children with cerebral palsy. *Developmental Medicine and Child Neurology*, 39(4), pp214–223. DOI: 10.1111/j.1469-8749.1997.tb07414.x. 14

Pauletto, S. and Hunt, A. (2006). The sonication of EMG data. *Proceedings of the International Conference on Auditory Display (ICAD)*, London, UK. 48, 51

Peretz, I. and Zatorre, R.J. (2005). Brain organization for music processing. *Annual Review of Psychology*, 56, pp. 89–114. DOI: 10.1146/annurev.psych.56.091103.070225. 51

Pigford, T. and Andrews, A.W. (2010) Feasibility and benefit of using the Nintendo Wii Fit for balance rehabilitation in an elderly patient experiencing recurrent falls. *Journal of Student Physical Therapy Research*, 2(1), pp. 12–20. 61

Powell L.E. and Myers, A.M. (1995) The activities-specific balance confidence (ABC) scale. *Journal of Gerontology*, 50A(1), M28–M34. DOI: 10.1093/gerona/50A.1.M28. 65

Rodda, J.M., Graham, H.K., Carson, L., Galea, M.P., and Wolfe, R. (2004) Sagittal gait patterns in spastic diplegia. *Journal of Bone and Joint Surgery*, 86(2), pp. 251–258. DOI: 10.1302/0301-620X.86B2.13878. 2

Rodger, M.W., Young, W.R., and Craig, C.M. (2014) Synthesis of walking sounds for alleviating gait disturbances in Parkinson's disease. *IEEE Transaction on Neural Systems and Rehabilitation Engineering*, 22(3), pp. 543–548. DOI: 10.1109/TNSRE.2013.2285410. 51

Romera-Paredes, B., Argyriou, A., Berthouze, N. ,and Pontil, M. (2012) Exploiting unrelated tasks in multi-task learning. *Proceedings of AISTATS '12, Conference on Artificial Intelligence and Statistics*, La Palma, Spain, pp. 951–959. 56

Romera-Paredes, B., Aung, H., Pontil, M., de C Williams, A.C., Watson, P., and Bianchi-Berthouze, N. (2013) Transfer learning to account for idiosyncrasy in face and body expressions. *Proceedings of FG2013, IEEE International Conference on Automatic Face and Gesture Recognition*, April 22nd-26th, Shanghai, China, pp. 1–6. DOI: 10.1109/fg.2013.6553779. 56

Rosati, G., Rodà, A., Avanzini, F., and Masiero, S. (2013) On the role of auditory feedback in robotic-assisted movement training after stroke. *Computational Intelligence and Neuroscience*, ID586138. Volume 2013, 15 pages DOI: 10.1155/2013/586138. 48, 51

Rose, D.J. (2008) Preventing falls among older adults: No "one size suits all" intervention strategy. *Journal of Rehabilitation Research and Development*, 45(8), pp. 1153–1166. DOI: 10.1682/JRRD.2008.03.0041. 59

Rosser, B.A., McCullagh, P., Davies, R., Mountain, G.A., McCracken, L., and Eccleston, C. (2011) Technology-mediated therapy for chronic pain management: the challenges of adapting behavior change interventions for delivery with pervasive communication technology. *Telemedicine Journal and e-health*, 17(3), pp. 211–216. DOI: 10.1089/tmj.2010.0136. 47

Rosser, B.A. and Ecclestone, C. (2011) Smart Phone Applications for Pain Management. *Journal of Telemedicine and Telecare*, 17(6), pp. 308–312. DOI: 10.1258/jtt.2011.101102. 47

Rudick, R.A., Miller, D., Bethoux, F., Rao, S.M., Lee, J.C., Stough, D., Reece, C., Schindler, D., Mamone, B., and Alberts, J. (2014) The multiple sclerosis performance test (MSPT): An iPad-based disability assessment tool. *Journal of Visualized Experiments*, (88), e51318. DOI: 10.3791/51318. 16

Ruppert, G., Amorim, P., Moares, T., and Silva, J. (2012) Touchless gesture user interface for 3D visualization using Kinect Platform and Open-Source Frameworks. *Proceedings of 5th International Conference on Advanced Research in Virtual and Rapid Prototyping*, September 28th – October 1st, Leiria, Portugal, pp. 215–219. 77

Salarian, A., Russmann, H., Vingerhoets, F.J., Dehollain, C., Blanc, Y., Burkhard, P.R., and Aminian, K. (2004) Gait assessment in Parkinson's disease: toward an ambulatory system for long-term monitoring. *IEEE Transactions on Biomedical Engineering*, 51(8), pp. 1434-43. DOI: 10.1109/TBME.2004.827933. 2

Sawhney, A., Agrawal, A., Patra, P., and Calvert, P. (2006) Piezoresistive sensors on textiles by inkjet printing and electroless plating. *Proceedings of Material Research Society Symposium on Smart Nanotextiles*, April 18th-19th, San Francisco, CA, pp. 103–112. DOI: 10.1557/proc-0920-s05-04. 6

Schaffert, N., Mattes, K., and Effenberg, A. O. (2010) Listen to the boat motion: acoustic information for elite rowers. *Proceedings of ISon 2010, 3rd Interactive Sonification Workshop, April 7th, Stockholm, Sweden, pp. 31-37.* 51

Schönauer, C., Pintaric, T., Kaufmann, H., Jansen Kosterink, S., and Vollenbroek-Hutten, M. (2011) Chronic pain rehabilitation with a serious game using multimodal input. *Proceedings of ICVR 2011, International Conference on Virtual Rehabilitation*, June 27th-29th, Zurich, Switzerland, pp. 1–8. DOI: 10.1109/ICVR.2011.5971855. 48

Scilingo, E.P., Lorussi, F., Mazzoldi, A., and de Rossi, D. (2003) Strain sensing fabrics for wearable kinaesthetic systems. *IEEE Sensors Journal*, 3(4), pp. 460–467. DOI: 10.1109/jsen.2003.815771. 6

Sergio, F., Singh, A., Bradley, C., Bianchi-Berthuouze, N., and de C Williams A. C. (2015) Roles for personal informatics in chronic pain. *Proceedings of PervasiveHealth '15*, May 20th–23rd 2015, Istanbul, Turkey, pp. 161–168. DOI: 10.4108/icst.pervasivehealth.2015.259501. 50

Sharara, H., Sopan, A., Namata, G., Getoor, L., and Singh, L. (2011) G-PARE: A visual analytic tool for comparative analysis of uncertain graphs. *Proceedings of VAST 2011, IEEE Conference on Visual Analytics Science and Technology*, Providence, RI, pp. 61–70. DOI: 10.1109/VAST.2011.6102442. 37

Sherrington, C., Tiedemann, A., Fairhall, N., Close, J.C., and Lord, S.R. (2011) Exercise to prevent falls in older adults: an updated meta-analysis and best practice recommendations. *NSW Public Health Bulletin*, 22(3-4), pp. 78–83. DOI: 10.1071/NB10056. 59

Shotton, J., Fitzgibbon, A., Cook, M., Sharp, T., Finocchio, M., Moore, R., Kipman, A., and A. Blake, A. (2011) Real-time human pose recognition in parts from single depth images. *Proceedings of CVPR '11, IEEE Conference on Computer Vision and Pattern Recognition*, 21st-23rd June, Colorado Springs, pp. 1297–1304. DOI: 10.1109/cvpr.2011.5995316. 5

Sihvonen, S.E., Sipilä, S., and Era, P.A. (2004) Changes in postural balance in frail elderly women during a 4-week visual feedback training: a randomized controlled trial. *Gerontology*, 50(2), pp. 87–95. DOI: 10.1159/000075559. 61

Simon, S.R. (2004) Quantification of human motion: gait analysis-benefits and limitations to its application to clinical problems. *Journal of Biomechanics* 37(12), pp. 1869–80. DOI: 10.1016/j.jbiomech.2004.02.047. 14

Simons, L.E., Elman, I., and Borsook, D. (2014) Psychological processing in chronic pain: A neural systems app. roach. *Neuroscience & Biobehavioral Reviews*, 39, pp. 61–78. DOI: 10.1016/j.neubiorev.2013.12.006. 47

Singh, A., Klapper, A., Jia, J., Fidalgo, A., Tajadura-Jimenez, A., Kanakam, N., Bianchi-Berthouze, N., and Williams, A. (2014) Motivating people with chronic pain to do physical activity: opportunities for technology design. *Proceedings of CHI '14, Conference on Human Factors in Computing Systems*, 26th April-1st May, Toronto, Canada, pp. 2803–2812. DOI: 10.1145/2556288.2557268. 2, 44, 49, 54

Singh, A., Piana, S., Pollarolo, D., Volpe, G., Varni, G., Tajadura-Jimenez, A., de C Williams, A., Camurri, A., and Bianchi-Berthouze, N. (2015) Go-with-the-flow: Tracking, analysis and sonification of movement and breathing to build confidence in activity despite chronic pain. *Human–Computer Interaction*. DOI: 10.1080/07370024.2015.1085310. 50, 54

Smeddinck, J., Herrlich, M., and Malaka, R. (2015) Exergames for physiotherapy and rehabilitation: A medium-term situated study of motivational aspects and impact on functional reach. *Proceedings of CHI '15 Conference on Human Factors in Computing Systems*, 18th-23rd April, Seoul, Korea, pp. 4143–4146. DOI: 10.1145/2702123.2702598. 44, 62

Smeddinck, J., Herrlich, M., Krause, M., Gerling, K., and Malaka, R. (2012) Did they really like the game? challenges in evaluating exergames with older adults. *Proceedings of the CHI 2012 Game User Research Workshop*, 5th-6th May, Austin, Texas. 62

Socie, M.J. and Sosnoff, J.J. (2013) Gait variability and multiple sclerosis. *Multiple Sclerosis International*, 2013:645197. DOI: 10.1155/2013/645197. 2

Spain, R.I., St George, R.J., Salarian, A., Mancini, M., Wagner, J.M., Horak, F.B., and Bourdette, D. (2012) Body-worn motion sensors detect balance and gait deficits in people with multiple sclerosis who have normal walking speed. *Gait & Posture*, 35(4), pp. 573–578. DOI: 10.1016/j.gaitpost.2011.11.026. 16

SPHERE Project. http://www.irc-sphere.ac.uk/.

Staub, C., Can, S., Knoll, A, Nitsch, V., Karl, I. ,and Farber, B. (2011) Implementation and evaluation of a gesture-based input method in robotic surgery. *Proceedings of HAVE 2011, IEEE International Workshop on Haptic Audio Visual Environments and Games*, October 14th-17th, Qinhuangdao, Hebei, China, pp.1–7. DOI: 10.1109/HAVE.2011.6088384. 92

Stern, H.I., Wachs, J.P., and Edan, Y. (2008) Optimal consensus intuitive hand gesture vocabulary design. *Proceedings of ICSC '08, IEEE International Conference on Semantic Computing*, August 4th-7th, Santa Clara, CA, pp. 96–103. DOI: 10.1109/icsc.2008.29. 76

Stokic, D.S., Horn, T.S., Ramshur, J.M., and Chow, J.W. (2009) Agreement between temporo-spatial gait parameters of an electronic walkway and a motion capture system in healthy and chronic stroke populations. *American Journal of Physical Medicine and Rehabilitation*, 88(6), pp. 437–444. DOI: 10.1097/PHM.0b013e3181a5b1ec. 2

Stone, E.E. and Skubic, M. (2011a) Evaluation of an inexpensive depth camera for passive in-home fall risk assessment. *Proceedings of PervasiveHealth Conference*, 23-26th May, Dublin, Ireland, pp. 71–77. DOI: 10.3233/AIS-2011-0124. 5

Stone, E.E. and Skubic, M. (2011b) Passive-in-home measurement of stride-to-stride gait variability comparing vision and Kinect sensing. *Proceedings of EMBC, '11,33rd Annual International Conference of the IEEE Engineering in Medicine and Biology Society*, August 30th – September 3rd, Boston, MA, pp. 6491–6494. DOI: 10.1109/iembs.2011.6091602. 5

Stone, E.E. and Skubic, M. (2013) Unobtrusive, continuous, in-home gait measurement using the Microsoft Kinect. *IEEE Transactions on Bio-Medical Engineering*, 60(10), pp. 2925–32. DOI: 10.1109/TBME.2013.2266341. 16, 42

Stratou, G., Scherer, S., Gratch, J., and Morency, L.P. (2014) Automatic nonverbal behavior indicators of depression and PTSD: the effect of gender. *Journal on Multimodal User Interfaces*, 9(1), pp. 17–29. DOI: 10.1007/s12193-014-0161-4.

Strickland, M., Tremaine, J., and Brigley, G. (2011) Team uses Xbox Kinect to see patient images during surgery. http://sunnybrook.ca/uploads/N110314.pdf.

Strickland, M., Tremaine, J., Brigley, G., and Law, C. (2013) Using a depth-sensing infrared camera system to access and manipulate medical imaging from within the sterile operating field. *Canadian Journal of Surgery*, 56(3), E1–6. DOI: 10.1503/cjs.035311. 76, 78, 79

Sullivan, M.J.L., Thibault, P., Savard, A., Catchlove, R., Kozey, J., and Stanish, W.D. (2006) The influence of communication goals and physical demands on different dimensions of pain behaviour. *Pain*, 125(3), pp. 270–277. DOI: 10.1016/j.pain.2006.06.019. 49

Sutherland, I. (1968) A head-mounted three dimensional display. *Proceedings of AFIPS '68, Fall Joint Computer Conference, part I*, December 9th-11th, San Francisco, CA, pp. 757–764. DOI: 10.1145/1476589.1476686.

Sweetser, P. and Wyeth, P. (2005) Gameflow: A model for evalutating player enjoyment in games. *Computers in Entertainment*, 3(3), p3. DOI: 10.1145/1077246.1077253.

Tajadura-Jiménez, A., Väljamäe, A., Toshima, I., Kimura, T., Tsakiris, M., and Kitagawa, N. (2012) Action sounds recalibrate perceived tactile distance. *Current Biology*, 22(13), R516–R517. DOI: h10.1016/j.cub.2012.04.028. 51

Takač, B., Catalá, A., Rodríguez Martín, D., van der Aa, N., Chen, W., and Rauterberg, M. (2013) Position and orientation tracking in a ubiquitous monitoring system for Parkinson disease patients with freezing of gait symptom. *Journal of Medical Internet Research: mHealth and uHealth*, 1(2), e14. DOI: 10.2196/mhealth.2539. 16, 42

Tan, J., Unal, J., Tucker, T., and Link, K. (2011) Kinect sensor for touch-free use in virtual medicine. http://www.youtube.com/watch?v=id7OZAbFaVI&feature=player_embedded. 77

Tang, Q., Vidrine, D., Crowder, E., and Intille, S. (2014a) Automated detection of puffing and smoking with wrist accelerometers. *Proceedings of PervasiveHealth '14*, Brussels, Belgium, Belgium, pp. 80–87. DOI: 10.4108/icst.pervasivehealth.2014.254978.

Tang, R., Alizadeh, H., Tang, A., Bateman, S., and Jorge. J. (2014b) Physio@Home: design explorations to support movement guidance. *Proceedings of CHI '14, Conference on Human Factors in Computing Systems*, April 26th-May 1st, Toronto, Canada, pp. 1651–1656. DOI: 10.1145/2559206.2581197. 44

Tao, W., Liu, T., Zheng, R., and Feng, H. (2012) Gait analysis using wearable sensors. *Sensors*, 12(2), pp. 2255–2283. DOI: 10.3390/s120202255.

Teber, D. Guven, S. Simpfendörfer, T., Baumhauer, M., Güven, E.O., Yencilek, F., Gözen, A.S., and Rassweiler, J. (2009) Augmented reality: A new tool to improve surgical accuracy during laparoscopic partial nephrectomy? Preliminary in vitro and in vivo results. *European Urology*, 56(2), pp. 332–338. DOI: 10.1016/j.eururo.2009.05.017.

Thelen, D.G., Muriuki, M., James, J., Schultz, A.B., Ashton-Miller, J.A., and Alexander, N.B. (2000) Muscle activities used by young and old adults when stepping to regain balance during a forward fall. *Journal of Electromyography and Kinesiology*, 10(2), pp. 93–101. DOI: 10.1016/S1050-6411(99)00028-0. 58

Thielgen, T., Foerster, F., Fuchs, G., Hornig, A., and Fahrenberg, J. (2004) Tremor in Parkinson's disease: 24-hr monitoring with calibrated accelerometry. *Electromyography and Clinical Neurophysiology*, 44(3), pp. 22–26. 16

Tien, C.L. (2009) Building Interactive Eyegaze Menus for Surgery. Master Thesis. Simon Fraser University, Canada.

Tracey, I. and Bushnell, M.C. (2009) How neuroimaging studies have challenged us to rethink: is chronic pain a disease? *The Journal of Pain*, 10(11), pp. 1113–1120. DOI: 10.1016/j.jpain.2009.09.001. 46

Tuncel, O., Altun, K., and Barshan, B. (2009) Classifying human leg motions with uniaxial piezoelectric gyroscopes. *Sensors*, 9(11), pp. 8508–8546. DOI: 10.3390/s91108508. 6

Turk, D.C. and Rudy, T.E. (1987) IASP taxonomy of chronic pain syndromes: preliminary assessment of reliability. *Pain*, 30(2), pp. 177–189. DOI: 10.1016/0304-3959(87)91073-6. 46

Ulbrecht, G., Wagner, D., and Gräßel, E. (2012) Exergames and their acceptance among nursing home residents. *Activities, Adaptation & Aging*, 36(2), pp. 93–106. DOI: 10.1080/01924788.2012.673155. 66

Usui, J., Hatayama, H., and Sato, T. (2006) Paravie: dance entertainment system for everyone to express oneself with movement. *Proceedings of ACE '06*, 14th-16th June, Hollywood, CA, article 30. DOI: 10.1145/1178823.1178861. 31

Uzor, S. and Baillie, L. (2013) Exploring & designing tools to enhance falls rehabilitation in the home. *Proceedings of CHI '13, Conference on Human Factors in Computing*, April 27th-May 2nd, Paris, France, pp. 1233–1242. DOI: 10.1145/2470654.2466159. 44, 59, 60, 61

Uzor, S. and Baillie, L. (2014) Investigating the long-term use of exergames in the home with elderly fallers. *Proceedings of CHI '14, Conference on Human Factors in Computing Systems*, April 26th-May 1st, Toronto, Canada, pp. 2813–2822. DOI: 10.1145/2556288.2557160. 44, 60

Uzor, S., Baillie, L., and Skelton, D. (2012) Senior designers: empowering seniors to design enjoyable falls rehabilitation tools (2012). *Proceedings of CHI '12*, May 5th-10th, Austin, Texas, pp. 1179–1188. DOI: 10.1145/2207676.2208568. 59

Valstar, M., Schuller, B., Smith, K., Almaev, T., Eyben, F., Krajewski, J., Cowie, R., and Pantic, M. (2014) AVEC 2014 - 3D dimensional affect and depression recognition challenge. *Proceedings of AVEC '14, 4th International Workshop on Audio/Visual Emotion Challenge*, November 7th, Orlando, FL, pp. 3–10. DOI: 10.1145/2661806.2661807.

van den Elzen, S. and van Wijk, J. J. (2011) BaobabView: Interactive construction and analysis of decision trees. *IEEE Conference on Visual Analytics Science and Technology (VAST)*, 2011, pp. 151–160. DOI: 10.1109/VAST.2011.6102453. 37

Vara-Thorbeck, C., Murioz, V. F., Toscano, R., Gomez, J., Fernandez, J., Felices, M., and Garcia-Cerezo, A. (2001) A new robotic endoscope manipulator a preliminary trial to evaluate the performance of a voice-operated industrial robot and a human assistant in several simulated and real endoscopic operations. *Surgical Endoscopy Journal of American Geriatric Society*, 15(9), pp. 924–927. DOI: h10.1007/s00464-001-0033-3. 92

Varni, G., Mancini, M., Volpe, G. and Camurri, A. (2011) A system for mobile active music listening based on social interaction and embodiment. *Mobile Network Application*, 16(3), pp. 375–384. DOI: 10.1007/s11036-010-0256-4.

Vaughan, C.L., Davis, B.L. and O'Connor, J. (1992) *Dynamics of Human Gait*. Human Kinetics Press, Champaign Illinois. 4

Vidulin, V., Bohanec, M., and Gams, M. (2014) Combining human analysis and machine data mining to obtain credible data relations. *Information Sciences*, 288, pp. 254–278. DOI: 10.1016/j.ins.2014.08.014. 37

Vidyarthi, J., Riecke, B.E., and Gromala, D. (2012) Sonic Cradle: designing for an immersive experience of meditation by connecting respiration to music. *Proceedings of DIS'12*, 11th-15th June, Newcastle, UK, pp. 408–417. DOI: 10.1145/2317956.2318017. 47, 51

Vogel, D. and Balakrishnan, R. (2005) Distant freehand pointing and clicking on very large, high resolution displays. *Proceedings of UIST '05, 18th Annual ACM symposium on User Interface Software and Technology*, 23rd-26th October, Seattle, WA, pp. 33–42. DOI: 10.1145/1095034.1095041. 87

Vogt, K., Pirro, D., Kobenz, I., Holdrich, R., and Eckel, G. (2009) PhysioSonic - evaluated movement sonification as auditory feedback in physiotherapy. *Proceedings of CMMR/ICAD*, 18th-22nd May, Copenhagen, Denmark, pp. 103–120. DOI: 10.1007/978-3-642-12439-6_6. 51

Volpe, della, R., Popa, T., Ginanneschi, F., Spidalieri, R., Mazzocchio, R., and Rossi, A. (2006) Changes in coordination of postural control during dynamic stance in chronic low back pain patients. *Gait & Posture* 24(3), pp. 349–355. DOI: 10.1016/j.gaitpost.2005.10.009. 47

Volpe, G. and Camurri, A. (2011) A system for embodied social active listening to sound and music content. *Journal on Computing and Cultural Heritage*, 4(1), Article 2. DOI: 10.1145/2001416.2001418.

Vorderer, P., Hartmann, T., and Klimmt, C. (2003) Explaining the enjoyment of playing video games: The role of competition. *Proceedings of ICEC '03, 2nd International Conference on Computer Games*, Pittsburgh, PA, pp. 1–9.

Wachs, J. (2009) Gaze, posture and gesture recognition to minimize focus shifts for intelligent operating rooms in a collaborative support system. *International Journal of Computers Communication and Control*, 5(1), pp. 106–124. DOI: 10.15837/ijccc.2010.1.2467. 90, 92

Wachs, J., Jacob, M., and Li, Y.T. (2012) Does a robotic scrub nurse improve economy of movements? *Proceedings of Medical Imaging 2012: Image-Guided Procedures, Robotic Interventions, and Modeling*, SPIE vol. 8316. DOI: 10.1117/12.911930. 89, 90

Wachs, J., Stern, H., Edan, Y., Gillam, M., Feied, C., Smith, M., and Handler, J. (2006) A real-time hand gesture interface for medical visualization applications. *Applications of Soft Computing*, 36, pp. 153–162. DOI: 10.1007/978-3-540-36266-1_15. 76

Watson, P.J., Booker, K.C., and Main, C.J. (1997) Evidence for the role of psychological factors in abnormal paraspinal activity in patients with chronic low back pain. *Journal of Musculoskeletal Pain*, 5(4), pp. 41–56. DOI: 10.1300/J094v05n04_05. 47

Weikert, M., Motl, R. W., Suh, Y., McAuley, E., and Wynn, D. (2010) Accelerometry in persons with multiple sclerosis: measurement of physical activity or walking mobility? *Journal of the Neurological Sciences*, 290(1-2), pp. 6–11. DOI: 10.1016/j.jns.2009.12.021. 16

Weiss, A., Sharifi, S., Plotnik, M., van Vugt, J.P., Giladi, N., and Hausdorff, J.M. (2011) Toward automated, at-home assessment of mobility among patients with Parkinson disease, using a body-worn accelerometer. *Neurorehabilitation and Neural Repair*, 25(9), pp. 810–818. DOI: 10.1177/1545968311424869. 16

Wellner, M., Schaufelberger, A., and Riener, R. (2007) A study on sound feedback in a virtual environment for gait rehabilitation. *Proceedings of 6th International Workshop on Virtual Rehabilitation*, Venice, Italy, pp. 53–56. DOI: 10.1109/icvr.2007.4362130. 48, 51

Westlake, K.P., Wu, Y.S., and Culham, E.G. (2007) Sensory-specific balance training in older adults: effect on proprioceptive reintegration. *Physical Therapy*, 87(10), pp. 1274–1283. DOI: 10.2522/ptj.20060263. 58

Whittle, M.W. (1996a) Clinical gait analysis: A review. *Human Movement Science*, 15, pp. 369–387. DOI: 10.1016/0167-9457(96)00006-1. 1

Whittle, M.W. (1996b) *Gait Analysis*. Oxford: Butterworth-Heinemann. 1

WHO (2002) *Innovative Care for Chronic Conditions: Building Blocks for Action*, World Health Organization, Geneva. 43

Whyatt, C., Merriman N.A., Young, W.R., Newell, F.N., and Craig, C.M. (2015) A Wii bit of fun: A novel platform to deliver effective balance training to older adults. *Games for Health Journal*, 4(6), pp. 423–33. DOI: 10.1089/g4h.2015.0006. 65, 66

Williams, B., Doherty, N.L., Bender, A., Mattox, H., and Tibbs, J.R. (2011) The effect of Nintendo Wii on balance: a pilot study supporting the use of the Wii in occupational therapy for the well elderly. *Occupational Therapy in Health Care*, 25(2-3), pp. 131–139. DOI: 10.3109/07380577.2011.560627. 61, 62

Wilson, R.N. (1958) Team work in the operating room. *Human Organization*, 12(4), pp. 9–14. DOI: 10.17730/humo.12.4.33300646734k2435.

Winter, D.A. (1995) Human balance and posture control during standing and walking. In *Gait & Posture*, 3(4), pp. 193–214. DOI: 10.1016/0966-6362(96)82849-9. 58, 61

Wolf, S.L., Barnhart, H.X., Ellison, G.L., and Coogler, C.E. (1997) The effect of Tai Chi Quan and computerized balance training on postural stability in older subjects. Atlanta FIC-SIT group. frailty and injuries: Cooperative studies on intervention techniques. *Physical Therapy*, 77(4), pp. 371–381. 61

Wolpert, D.M. and Ghahramani, Z. (2000) Computational principles of movement neuroscience. *Nature Neuroscience*, 3, pp. 1212–1217. DOI: 10.1038/81497. 51

Wong, W.M., Wong, M.S., and Lo, K.H. (2007) Clinical applications of sensors for human posture and movement analysis: A review. *Prosthetics and Orthotics International*, 31(1), pp. 62–75. DOI: 10.1080/03093640600983949. 6

Wulf, G. and Prinz, W. (2001) Directing attention to movement effects enhances learning: A review. *Psychonomic Bulletin & Review*, 8(4), pp. 648–660. DOI: 10.3758/BF03196201. 61

Yaniv, Z. (2008) Rigid registration. In Peters, T.M. and Cleary, K. (Eds) *Image-Guided Interventions: Technology and Applications*. Heidelberg, Germany: Springer, pp. 159–192. DOI: 10.1007/978-0-387-73858-1_6.

Young, W., Ferguson, S., Brault, S., and Craig, C. (2011) Assessing and training standing balance in older adults: a novel approach using the 'Nintendo Wii' Balance Board. *Gait & Posture* 33(2), pp. 303–305. DOI: 10.1016/j.gaitpost.2010.10.089. 16, 62, 66

Young, W.R., Rodger, M.W.M., and Craig, C.M. (2014). Auditory observation of stepping actions can cue both spatial and temporal components of gait in Parkinson's disease patients. *Neuropsychologia*, 57, pp. 140–153. DOI: 10.1016/j.neuropsychologia.2014.03.009. 51

Author Biographies

Prof. Kenton O'Hara works in the Human Experience and Design Group at Microsoft Research and is a Visiting Professor in the Computer Science Department at the University of Bristol. His research explores everyday social and collaborative practices with technology with a view to informing design and innovation. His most recent research has focused on user experiences and practices with "touchless" gestural interaction technology with a particular emphasis on its application in surgery. Over the years, his research has investigated new technologies in a variety of domains including the home, mobile environments, urban settings, and the workplace. Kenton has authored over 100 publications and two books on public displays and music consumption. Prior to working for Microsoft Research, Kenton worked as a Senior Principal Scientist at the Commonwealth Scientific and Industrial Research Organisation in Australia where he was Director of the HxI Initiative. He also worked as a Senior Researcher at Xerox EuroPARC, Hewlett-Packard Laboratories, and the Appliance Studio. He has worked on numerous award-winning projects including the BBC's BAFTA and Royal Television Society award winning "Coast" location-based experience.

Dr. Cecily Morrison is a researcher at Microsoft Research in the Human Experience & Design group. Cecily works is the area of Human-Computer Interaction, holding a B.A. in Anthropology from Columbia University and a Ph.D. in Computer Science from the University of Cambridge. Her interests lie in developing novel technologies to enable health and well-being in the broadest sense. Recent work has been focused on the integration of computer vision, and machine intelligence more generally, into useful real-world applications, such as the Assess MS system to support disease tracking in patients with Multiple Sclerosis. A key tenant of Cecily's research is to symbiotically bring together people and systems so that systems augment rather than replace what people do. Much of Cecily's inspiration came from years of working with health professionals to "translate" technology research into systems that could be used in practice.

Prof. Abigail Sellen is a Principal Researcher at Microsoft Research, UK, where she manages the Human Experience & Design Group, a group concerned with human experiences with computing, drawing on diverse perspectives across the sciences, engineering, arts, and humanities. Designing systems which use computer vision and machine learning for real-world situations, particularly in the area of health and well-being, is a major research theme for the group. Prior to joining Microsoft, Abigail worked at Hewlett Packard Labs, Bristol, Xerox's research lab in Cambridge UK (EuroPARC), the MRC Applied Psychology Unit, Cambridge, and other corporate IT labs such as Xerox PARC, Apple Computer, and Bell Northern Research. She has a doctorate in Cognitive Science from the University of California, San Diego and an M.A.Sc. in Industrial Engineering from the University of Toronto. She has published extensively on many topics including: computer input, help systems, reading, paper use in offices, videoconferencing design, search, photo use, gesture-based input, human error and computer support for human memory. This includes the book *The Myth of the Paperless Office* (with co-author Richard Harper), which won an IEEE award. She is a Fellow of the Royal Academy of Engineering, a Fellow of the British Computer Society, a Fellow of the Women's Engineering Society, an Honorary Professor of Interaction at the University of Nottingham, an Honorary Professor at UCLIC, University College London, and a member of the ACM SIGCHI Academy.

Prof. Nadia Bianchi-Berthouze is a Full Professor in Affective Computing and Interaction at the Interaction Centre of the University College London (UCL). She received her Ph.D. in Computer Science for Biomedicine from the University of the Studies of Milan, Italy. Her research focuses on designing technology that can sense the affective state of its users and use that information to tailor the interaction process. She has pioneered the field of Affective Computing and for more than a decade she has investigated body movement and more recently touch behavior as means to recognize and measure the quality of the user experience in full-body computer games, physical rehabilitation, and textile design. She also studies how full-body technology and body sensory feedback can be used to modulate people's perception of themselves and of their capabilities to improve self-efficacy and copying capabilities. She has published more than 170 papers in Affective Computing, HCI, and Pattern Recognition.

She was awarded the 2003 Technical Prize from the Japanese Society of Kansei Engineering and she has given a TEDxStMartin talk (2012).

Prof. Cathy Craig is a professor of Perception and Action Psychology and Director of the state-of-the-art Movement Innovation Lab at Queen's University Belfast, N. Ireland. Her cutting-edge research is primarily concerned with how sensory information picked up by the brain is subsequently used to guide all kinds of action. She uses this knowledge to develop innovative interventions that help improve movement performance in different groups of people (e.g. older adults, people with Parkinson's disease, children with autism). The caliber of her work has been recognized through the award of a prestigious ERC (European Research Council) grant, reserved for the very best scientists in Europe. This funding has allowed her to test how different types of multi-sensory virtual environments can be used to invite users to move in certain ways, whilst taking into account the action capabilities of the end-user. By adopting some basic principles of perception/action coupling she has developed a series of gamified scenarios that intrinsically motivate older adults to perform certain actions. To date she has successfully applied her perception/action research to successfully improve balance control and mobility levels in older adults, but also to understand differences in performance in elite sportsmen and women. Her ERC funded work was one of only four projects chosen to be showcased at the 1st Innovation EU Convention in Brussels and the ERC's 5th Birthday celebrations. She was runner up in the Health Innovation Awards in N. Ireland in 2011 and was a finalist in the 25k Innovation awards for her work on Parkinson's and falls prevention in older adults.

Printed in the United States
by Baker & Taylor Publisher Services